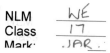

D1628271

The Concise Book of Muscles

second edition

Chris Jarmey

Lotus Publishing
Chichester, England

North Atlantic Books
Berkeley, California

First published in 2003. This second edition published in 2008 by
Lotus Publishing
Apple Tree Cottage, Inlands Road, Nutbourne, Chichester, PO18 8RJ and
North Atlantic Books
PO Box 12327
Berkeley, California 94712

Anatomical Drawings Amanda Williams
Line Drawings Chris Fulcher
Text Design Wendy Craig
Cover Design Paula Morrison
Printed and Bound in the UK by Scotprint

The Concise Book of Muscles is sponsored by the Society for the Study of Native Arts and Sciences, a nonprofit educational corporation whose goals are to develop an educational and crosscultural perspective linking various scientific, social, and artistic fields; to nurture a holistic view of arts, sciences, humanities, and healing; and to publish and distribute literature on the relationship of mind, body, and nature.

British Library Cataloguing in Publication Data
A CIP record for this book is available from the British Library
ISBN 978 1 905367 11 5 (Lotus Publishing)
ISBN 978 1 55643 719 9 (North Atlantic Books)

The Library of Congress has cataloged the first edition as follows:

Jarmey, Chris.
The concise book of muscles / Chris Jarmey.
p. ; cm. ill.
Includes bibliographical references and index.
 ISBN 1-55643-466-9 (pbk. : alk. paper)
1. Muscles--Handbooks, manuals, etc.
 [DNLM: 1. Muscles--innervation--Atlases. 2. Muscles--innervation--Handbooks. 3. Anatomy, Regional--Atlases. 4. Anatomy, Regional--Handbooks. 5. Muscles--anatomy & histology--Atlases.
6. Muscles--anatomy & histology--Handbooks. WE 17 J37c 2003] I. Title.
QP321.J43 2003
612.7'4--dc21
2002155580

Contents

About this Book

This book is designed in quick reference format to offer useful information about the main skeletal muscles that are central to sport, dance and exercise. Each muscle section is colour-coded for ease of reference. Enough detail is included regarding each muscle's origin, insertion and action commensurate with the requirements of the student and practitioner of bodywork, movement therapies and the movement arts. It aims to present that information accurately, in a particularly clear and user-friendly format; especially as anatomy can seem heavily laden with technical terminology. Technical terms are therefore explained in parenthesis throughout the text. The information about each muscle is presented in a uniform style throughout. An example is given below, with the meaning of headings explained in bold (some muscles will have abbreviated versions of this).

The attachment that remains relatively fixed during muscular contraction, i.e. the end of the muscle that is fixed to the bone that does not move, thereby acting as an anchor for the muscle to pull its opposite end (insertion) towards this fixed attachment (see p.28)

The attachment that moves (i.e. at the opposite end of the muscle to the origin). Note that when the insertion remains relatively fixed and the origin moves, the muscle is said to be causing a 'reverse origin to insertion' movement. This occurs often. Generally, the origin is more proximal (towards the centre of the body) and the insertion is more distal (towards the periphery of the body)

The name of the muscle

Some fundamental exercises to strengthen the muscle

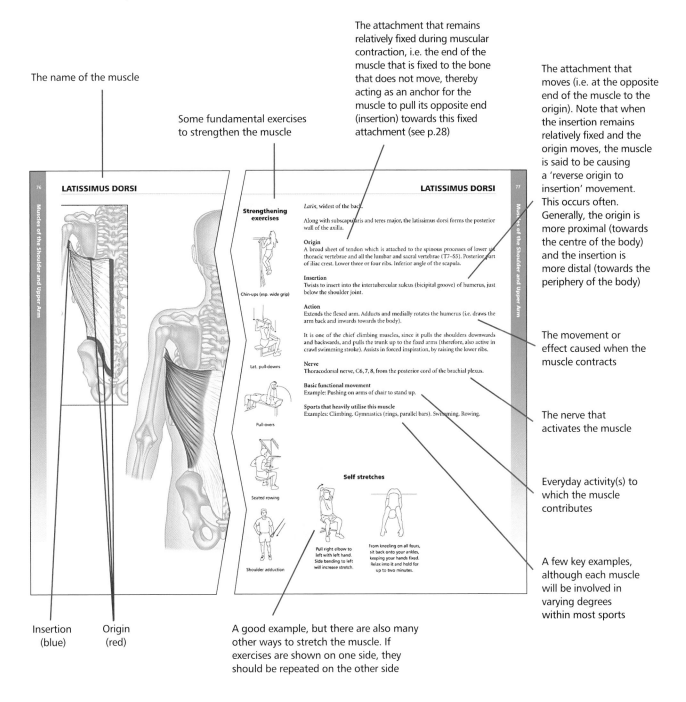

The movement or effect caused when the muscle contracts

The nerve that activates the muscle

Everyday activity(s) to which the muscle contributes

A few key examples, although each muscle will be involved in varying degrees within most sports

Insertion (blue)   Origin (red)

A good example, but there are also many other ways to stretch the muscle. If exercises are shown on one side, they should be repeated on the other side

A Note About Peripheral Nerve Supply

The nervous system comprises:

- The central nervous system (i.e. the brain and spinal cord).
- The peripheral nervous system (including the autonomic nervous system, i.e. all neural structures outside the brain and spinal cord).

The peripheral nervous system consists of 12 pairs of cranial nerves and 31 pairs of spinal nerves (with their subsequent branches). The spinal nerves are numbered according to the level of the spinal cord from which they arise (the level is known as the *spinal segment*).

The relevant peripheral nerve supply is listed with each muscle presented in this book, for those who need to know. However, information about the spinal segment* from which the nerve fibres emanate often differs between the various sources. This is because it is extremely difficult for anatomists to trace the route of an individual nerve fibre through the intertwining maze of other nerve fibres as it passes through its plexus (plexus = a network of nerves: from the Latin word meaning 'braid'). Therefore, such information has been derived mainly from empirical clinical observation, rather than through dissection of the body.

In order to give the most accurate information possible, I have duplicated the method devised by Florence Peterson Kendall and Elizabeth Kendall McCreary (*see* resources: Muscles Testing and Function). Kendall & McCreary integrated information from six well-known anatomy reference texts; namely, those written by: Cunningham, deJong, Foerster & Bumke, Gray, Haymaker & Woodhall, and Spalteholz. Following the same procedure, and then cross-matching the results with those of Kendall & McCreary, the following system of emphasising the most important nerve roots for each muscle has been adopted in this book.

Let us take the supinator muscle as our example, which is supplied by the deep radial nerve, C5, **6**, (7). The relevant spinal segment is indicated by the letter [C] and the numbers [5, **6**, (7)]. Bold numbers [e.g. **6**] indicate that most (at least five) of the sources agree. Numbers that are not bold [e.g. 5] reflect agreement by three of four sources. Numbers not in bold and in parenthesis [e.g. (7)] reflect agreement by two sources only, or if more than two sources specifically regarded it as a very minimal supply. If a spinal segment was mentioned by only one source, it was disregarded. Hence, bold type indicates the major innervation; not bold indicates the minor innervation; and numbers in parenthesis suggest possible or infrequent innervation.

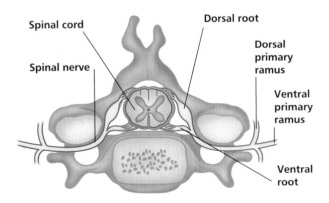

Figure 1: A spinal segment, showing the nerve roots combining to form a spinal nerve, which then divides into ventral and dorsal rami.

* A spinal segment is the part of the spinal cord that gives rise to each pair of spinal nerves (a pair consists of one nerve for the left side and one for the right side of the body). Each spinal nerve contains motor and sensory fibres. Soon after the spinal nerve exits through the foramen (the opening between adjacent vertebrae), it divides into a dorsal primary ramus (directed posteriorly) and a ventral primary ramus (directed laterally or anteriorly). Fibres from the dorsal rami innervate the skin and extensor muscles of the neck and trunk. The ventral rami supply the limbs, plus the sides and front of the trunk.

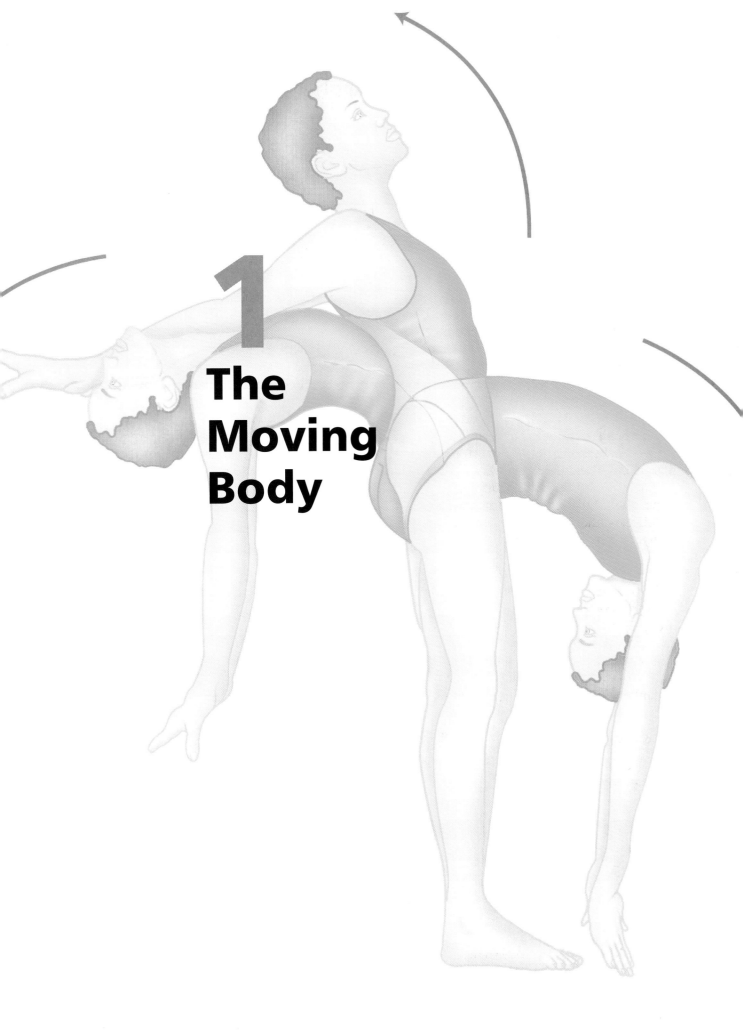

1 The Moving Body

Anatomical Directions

To describe the relative position of body parts and their movements, it is essential to have a universally accepted initial reference position. The standard body position known as the anatomical position serves as this reference. The *anatomical position* is simply the upright standing position with arms hanging by the sides, palms facing forwards (*see* figure 2). Most directional terminology used refers to the body *as if* it were in the anatomical position, regardless of its actual position. Note also that the terms 'left' or 'right' refer to the sides of the object or person being viewed, and not those of the reader.

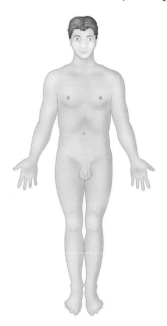

Figure 2: **Anterior.**
In front of; toward or at the front of the body.

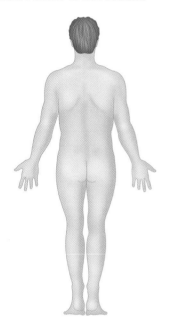

Figure 3: **Posterior.**
Behind; toward or at the backside of the body.

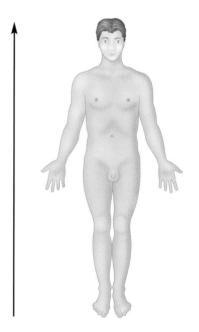

Figure 4: **Superior.**
Above; toward the head or upper part
of the structure or the body.

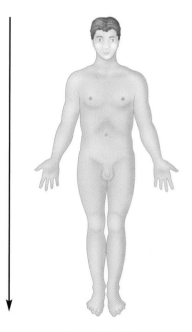

Figure 5: **Inferior.**
Below; away from the head or toward the
lower part of a structure or the body.

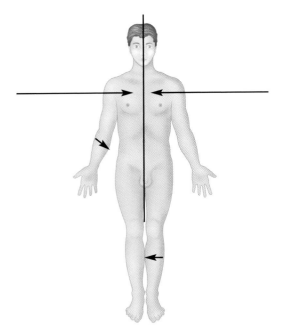

Figure 6: **Medial.**
(from *medius* in Latin, meaning middle)
Toward or at the midline of the body;
on the inner side of a limb.

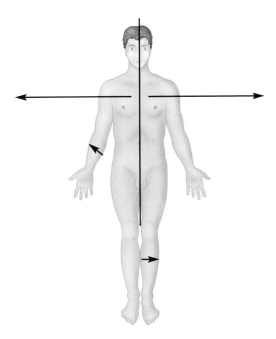

Figure 7: **Lateral.**
(from *latus* in Latin, meaning side)
Away from the midline of the body;
on the outer side of the body or a limb.

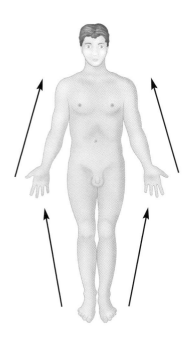

Figure 8: **Proximal.**
(from *proximus* in Latin, meaning next to)
Closer to the centre of the body (the navel), or to the point
of attachment of a limb to the body torso.

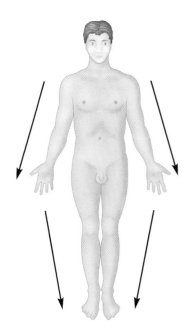

Figure 9: **Distal.**
(from *distans* in Latin, meaning distant)
Farther from the centre of the body, or from the point
of attachment of a limb to the torso.

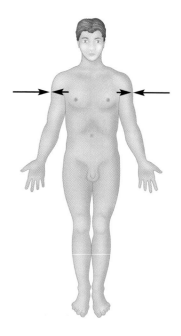

Figure 10: **Superficial.**
Toward or at the body surface.

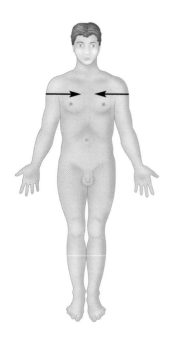

Figure 11: **Deep.**
Farther away from the body surface;
more internal.

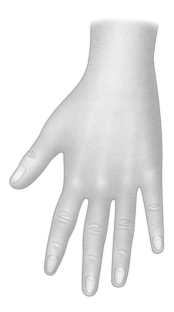

Figure 12: **Dorsum.**
The posterior surface of something,
e.g. the back of the hand;
the top of the foot.

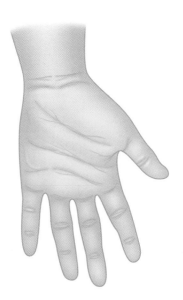

Figure 13: **Palmar.**
The anterior surface of the hand,
i.e. the palm.

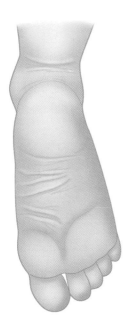

Figure 14: **Plantar.**
The sole of the foot.

Regional Areas

The two primary divisions of the body are its *axial* part, consisting of the head, neck and trunk, and its *appendicular* parts, consisting of the limbs that are attached to the axis of the body. Figure 15 shows the terms used to indicate specific body areas. Terms enclosed within brackets refer to the lay term for the area.

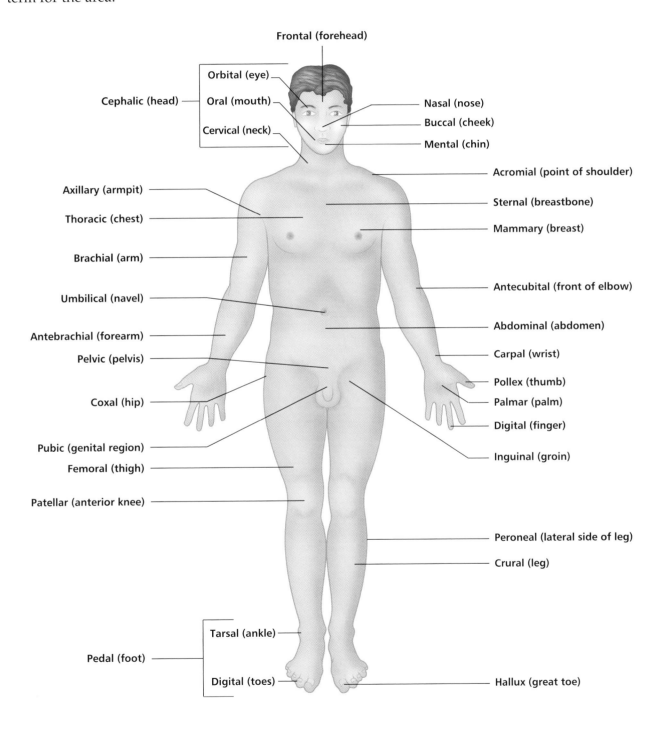

Frontal (forehead)

Orbital (eye)

Cephalic (head)

Oral (mouth)

Cervical (neck)

Nasal (nose)

Buccal (cheek)

Mental (chin)

Acromial (point of shoulder)

Axillary (armpit)

Sternal (breastbone)

Thoracic (chest)

Mammary (breast)

Brachial (arm)

Antecubital (front of elbow)

Umbilical (navel)

Abdominal (abdomen)

Antebrachial (forearm)

Carpal (wrist)

Pelvic (pelvis)

Pollex (thumb)

Coxal (hip)

Palmar (palm)

Digital (finger)

Pubic (genital region)

Inguinal (groin)

Femoral (thigh)

Patellar (anterior knee)

Peroneal (lateral side of leg)

Crural (leg)

Tarsal (ankle)

Pedal (foot)

Digital (toes)

Hallux (great toe)

Figure 15: Terms used to indicate specific body areas;
a) anterior view.

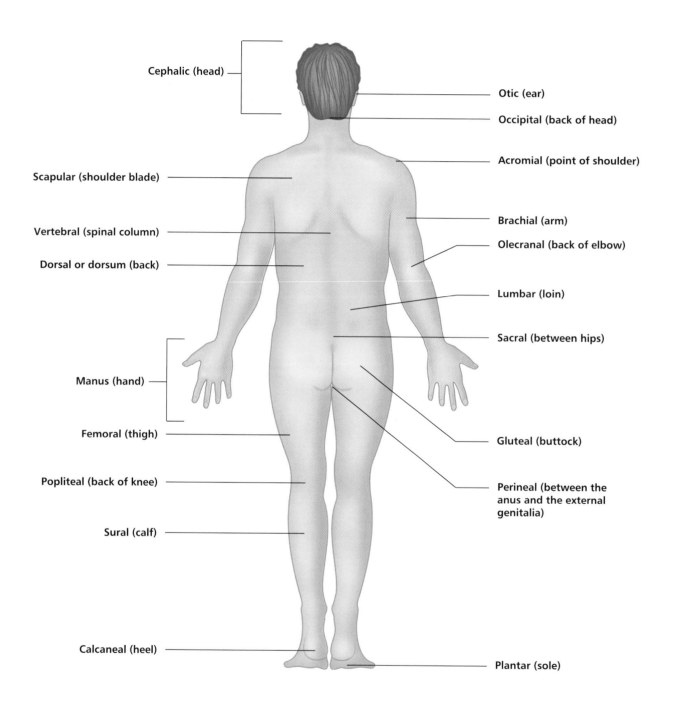

Cephalic (head)

Otic (ear)

Occipital (back of head)

Acromial (point of shoulder)

Scapular (shoulder blade)

Brachial (arm)

Vertebral (spinal column)

Olecranal (back of elbow)

Dorsal or dorsum (back)

Lumbar (loin)

Sacral (between hips)

Manus (hand)

Femoral (thigh)

Gluteal (buttock)

Popliteal (back of knee)

Perineal (between the
anus and the external
genitalia)

Sural (calf)

Calcaneal (heel)

Plantar (sole)

Figure 15: Terms used to indicate specific body areas;
b) posterior view.

Planes of the Body

Planes refer to two-dimensional sections through the body, to give a view of the body or body part, as if it has been cut through an imaginary line.

- The sagittal planes cut vertically through the body from anterior to posterior, dividing the body into right and left halves. The illustration shows the mid-sagittal plane.
- The frontal (coronal) planes pass vertically through the body, dividing the body into anterior and posterior sections, and lie at right angles to the sagittal plane.
- The transverse planes are horizontal cross sections, dividing the body into upper (superior) and lower (inferior) sections, and lie at right angles to the other two planes. Figure 16 illustrates the most frequently used planes.

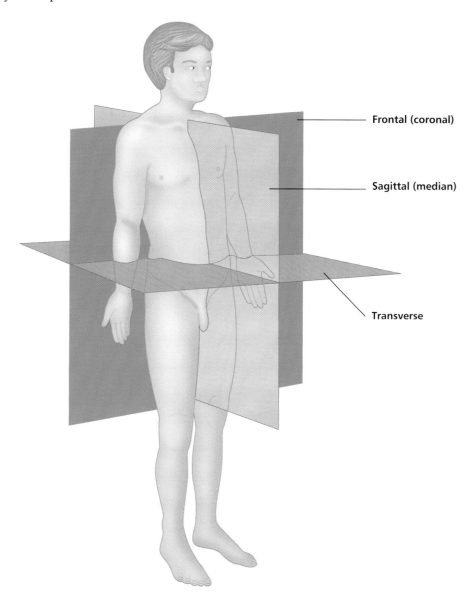

Figure 16: Planes of the body.

Anatomical Movements

The direction that body parts move is described in relation to the foetal (fetal) position. Moving into the foetal position results from flexion of all the limbs. Straightening out of the foetal position results from extension of all the limbs.

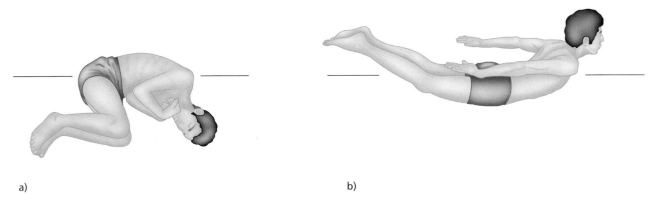

a)   b)

Figure 17: a) Flexion into the foetal position; b) extension out of the foetal position.

Main Movements

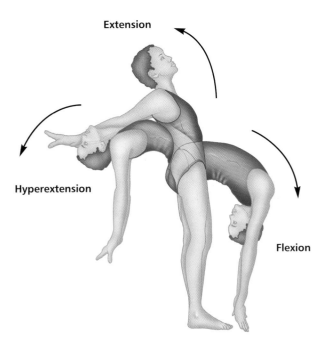

Figure 18: **Flexion:** Bending to decrease the angle between bones at a joint. From the anatomical position, flexion is usually forward, except at the knee joint where it is backward. The way to remember this is that flexion is always toward the foetal position.
Extension: To straighten or bend backward away from the foetal position.
Hyperextension: To extend the limb beyond its normal range.

Figure 19: **Lateral flexion.**
To bend the torso or head laterally (sideways) in the frontal (coronal) plane.

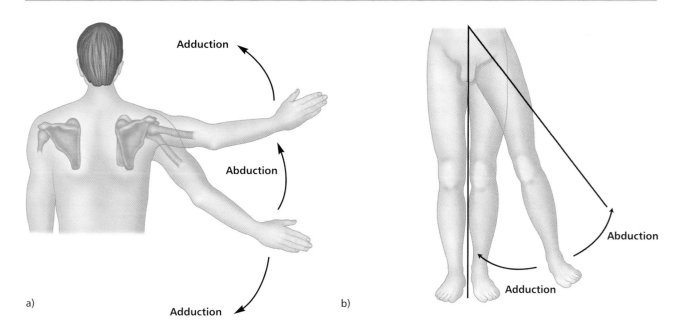

a)   b)

Figure 20: **Abduction:** Movement of a bone away from the midline of the body or the midline of a limb.
Adduction: Movement of a bone towards the midline of the body or the midline of a limb.

NOTE: for abduction of the arm to continue above the height of the shoulder (elevation through abduction, *see* page 19), the scapula must rotate on its axis to turn the glenoid cavity upwards (*see* figure 28b).

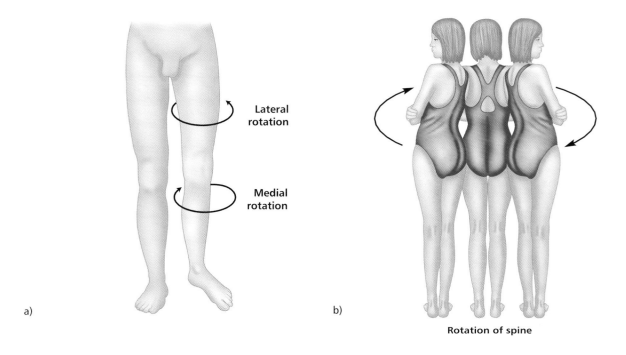

a)   b)

Rotation of spine

Figure 21:
Rotation: Movement of a bone or the trunk around its own longitudinal axis.
Medial rotation: to turn in towards the midline.
Lateral rotation: to turn out, away from the midline.

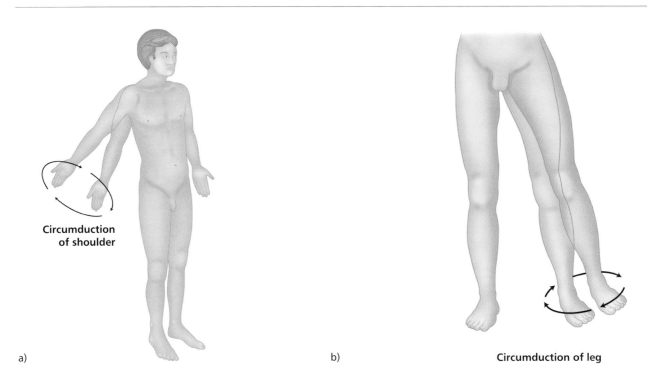

a)   b)   **Circumduction of leg**

Circumduction
of shoulder

Figure 22: **Circumduction.**
Movement in which the distal end of a bone moves in a circle, while the proximal end remains stable;
the movement combines flexion, abduction, extension, and adduction.

Other Movements

Movements in this section are those that occur only at specific joints or parts of the body, usually involving more than one joint.

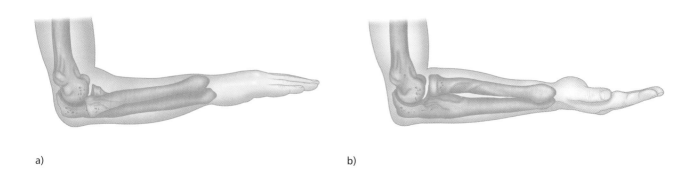

a)   b)

Figure 23a: **Pronation.**
To turn the palm of the hand down to face the floor
(if standing with elbow bent 90°, or if lying flat on the floor),
or away from the anatomical and foetal positions.

Figure 23b: **Supination.**
To turn the palm of the hand up to face the ceiling
(if standing with elbow bent 90°, or if lying flat on the floor),
or toward the anatomical and foetal positions.

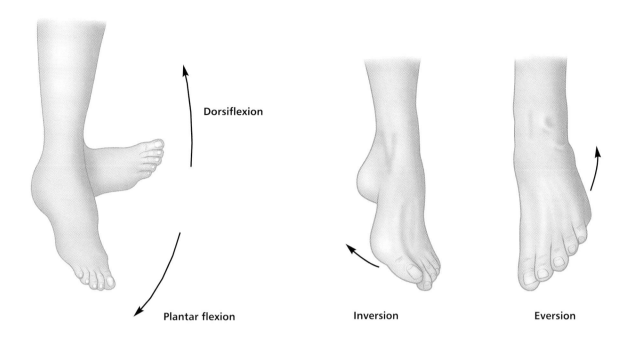

Figure 24: **Plantar flexion:** To point the toes down towards the ground. **Dorsiflexion**: To point the toe towards the sky.

Figure 25: **Inversion:** To turn the sole of the foot inward, so that the soles would face towards each other. **Eversion:** To turn the sole of the foot outward, so that the soles would face away from each other.

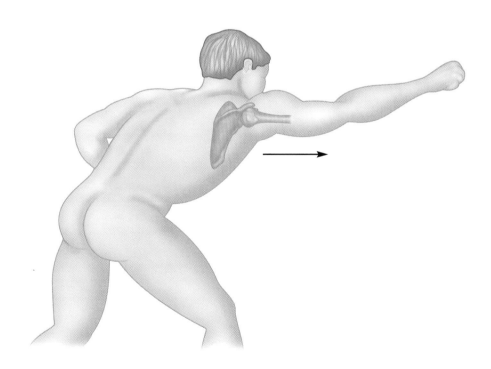

Figure 26: **Protraction.**
Movement forwards in the transverse plane.
For example, protraction of the shoulder girdle, as in rounding the shoulder.

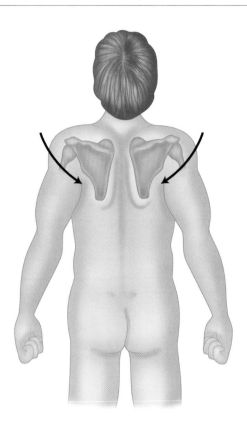

Figure 27: **Retraction.**
Movement backward in the transverse plane,
as in bracing the shoulder girdle back, military style.

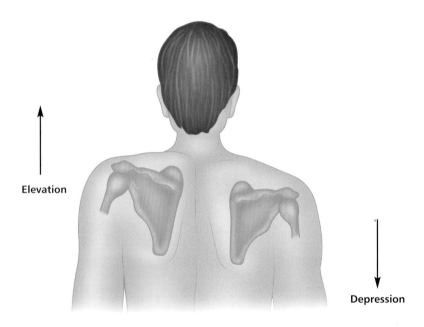

Figure 28a:
Elevation: Movement of a part of the body upwards along the frontal plane.
For example, elevating the scapula by shrugging the shoulders.
Depression: Movement of an elevated part of the body downward to its original position.

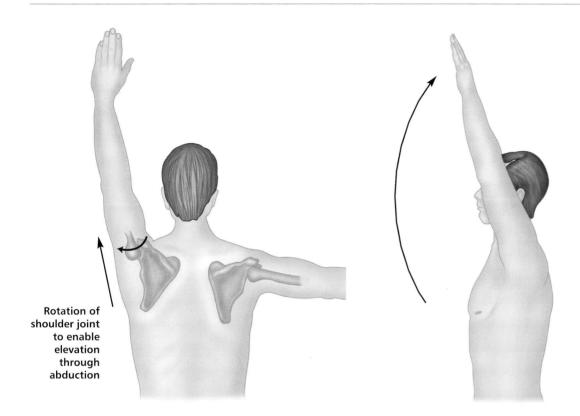

Rotation of shoulder joint to enable elevation through abduction

Figure 28b: Abducting the arm at the shoulder joint, then continuing to raise it above the head in the frontal plane can be referred to as **elevation through abduction.**

Figure 28c: Flexing the arm at the shoulder joint, then continuing to raise it above the head in the sagittal plane can be referred to as **elevation through flexion.**

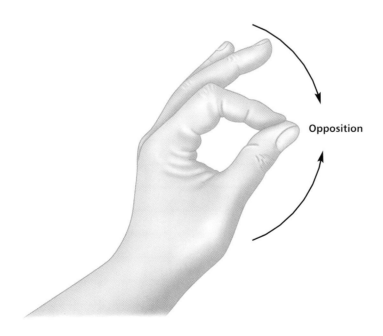

Opposition

Figure 29: **Opposition.**
A movement specific to the saddle joint of the thumb, that enables you to touch your thumb to the tips of the fingers of the same hand.

The Skeletal System

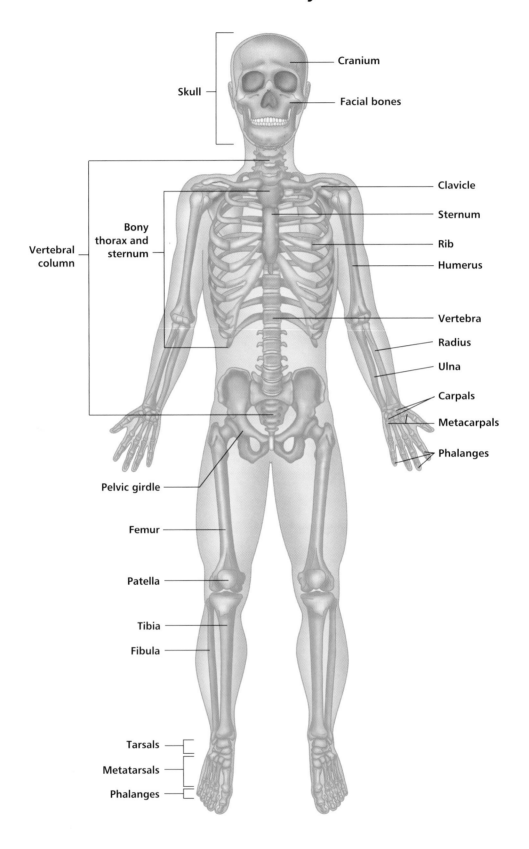

Figure 30a: Skeleton (anterior view).

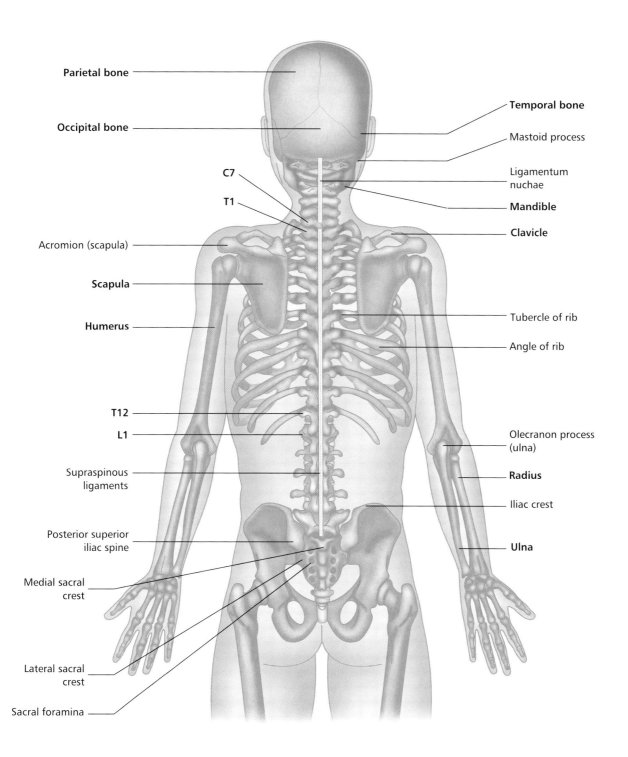

Parietal bone

Occipital bone

C7

T1

Acromion (scapula)

Scapula

Humerus

T12

L1

Supraspinous
ligaments

Posterior superior
iliac spine

Medial sacral
crest

Lateral sacral
crest

Sacral foramina

Temporal bone

Mastoid process

Ligamentum
nuchae

Mandible

Clavicle

Tubercle of rib

Angle of rib

Olecranon process
(ulna)

Radius

Iliac crest

Ulna

Figure 30b: Skeleton (posterior view).

Sections of the Vertebral Column

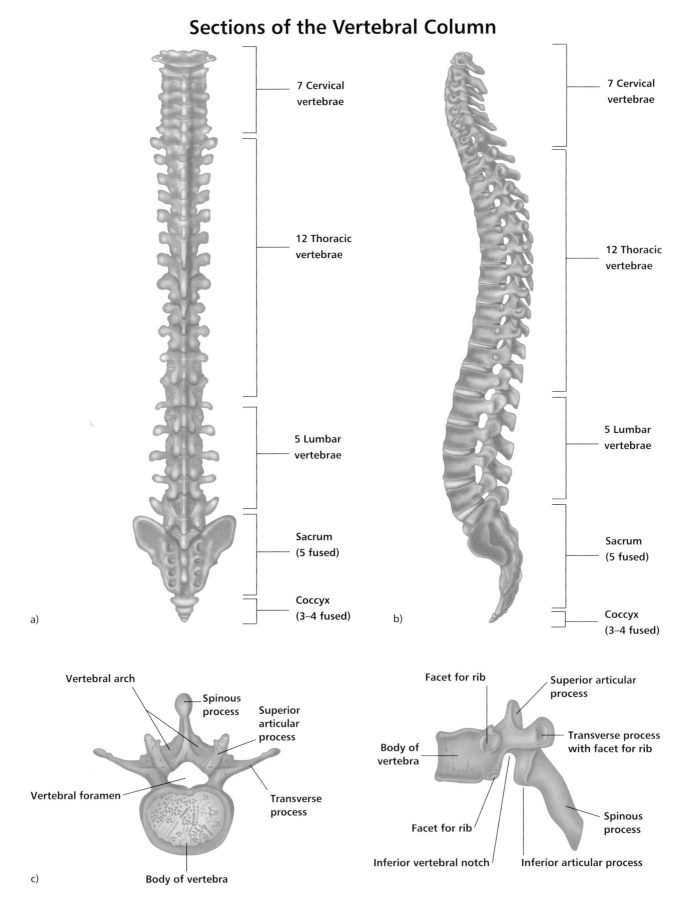

Figure 31: a) posterior view, b) lateral view, and c) vertebrae: lumbar (superior view) and thoracic (lateral view).

Thoracic to Pelvic Region

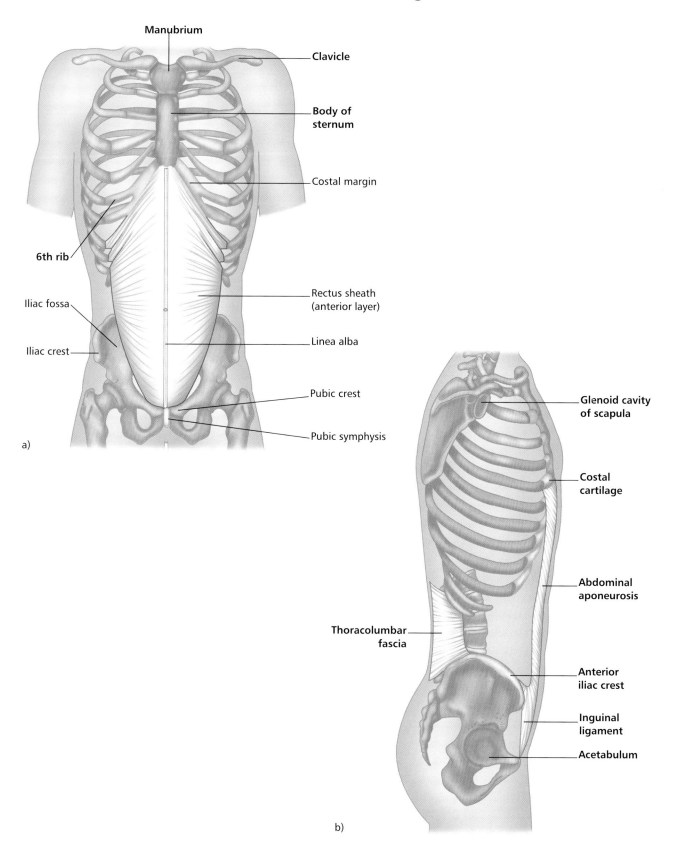

Figure 32: a) anterior view, and b) lateral view.

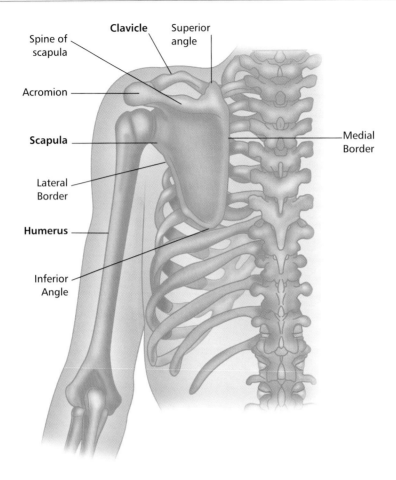

Figure 33: The scapula (posterior view).

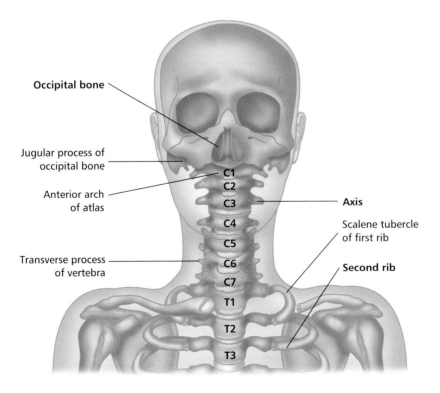

Figure 34: Skull to sternum (anterior view, the mandible and maxilla are removed).

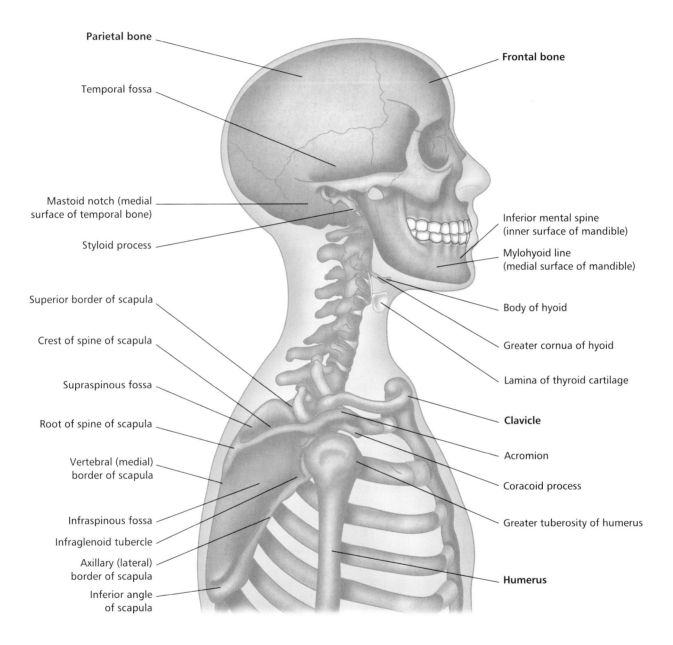

Parietal bone

Temporal fossa

Mastoid notch (medial
surface of temporal bone)

Styloid process

Superior border of scapula

Crest of spine of scapula

Supraspinous fossa

Root of spine of scapula

Vertebral (medial)
border of scapula

Infraspinous fossa

Infraglenoid tubercle

Axillary (lateral)
border of scapula

Inferior angle
of scapula

Frontal bone

Inferior mental spine
(inner surface of mandible)

Mylohyoid line
(medial surface of mandible)

Body of hyoid

Greater cornua of hyoid

Lamina of thyroid cartilage

Clavicle

Acromion

Coracoid process

Greater tuberosity of humerus

Humerus

Figure 35: Skull to humerus (lateral view).

The Muscular System

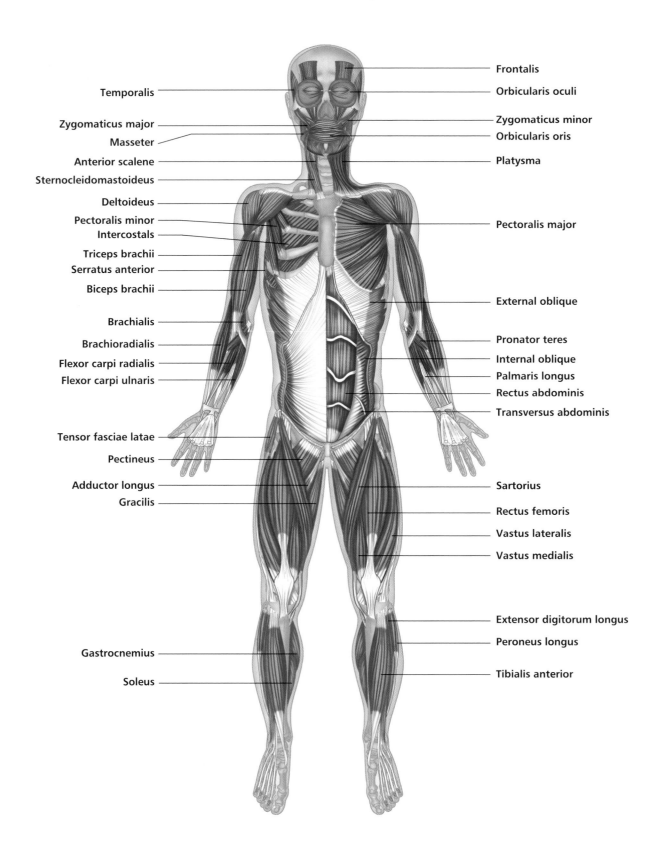

Temporalis

Zygomaticus major
Masseter
Anterior scalene
Sternocleidomastoideus
Deltoideus
Pectoralis minor
Intercostals
Triceps brachii
Serratus anterior
Biceps brachii

Brachialis
Brachioradialis
Flexor carpi radialis
Flexor carpi ulnaris

Tensor fasciae latae
Pectineus
Adductor longus
Gracilis

Gastrocnemius
Soleus

Frontalis
Orbicularis oculi
Zygomaticus minor
Orbicularis oris
Platysma

Pectoralis major

External oblique

Pronator teres
Internal oblique
Palmaris longus
Rectus abdominis
Transversus abdominis

Sartorius
Rectus femoris
Vastus lateralis
Vastus medialis

Extensor digitorum longus
Peroneus longus
Tibialis anterior

Figure 36a: Muscular system (anterior view).

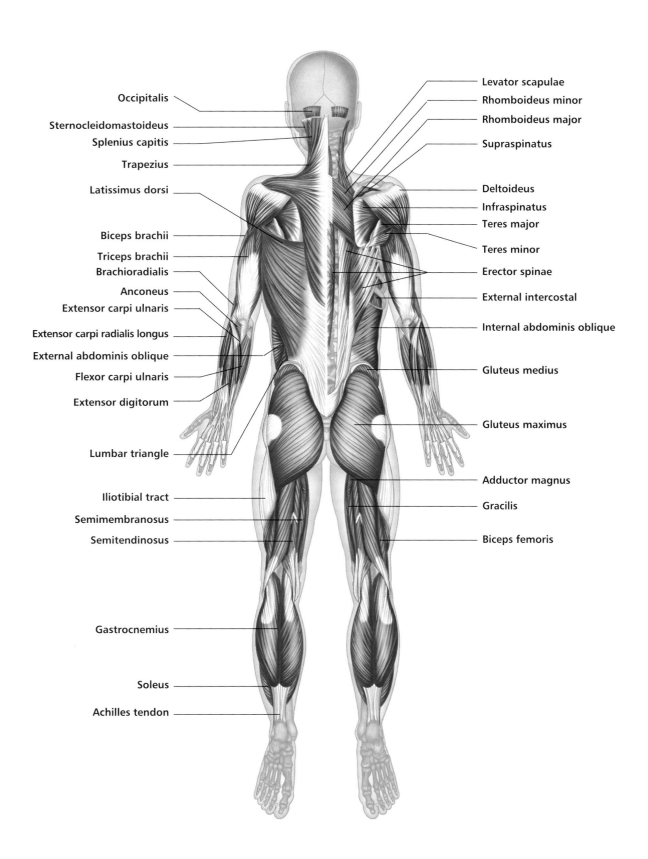

Occipitalis

Sternocleidomastoideus

Splenius capitis

Trapezius

Latissimus dorsi

Biceps brachii

Triceps brachii

Brachioradialis

Anconeus

Extensor carpi ulnaris

Extensor carpi radialis longus

External abdominis oblique

Flexor carpi ulnaris

Extensor digitorum

Lumbar triangle

Iliotibial tract

Semimembranosus

Semitendinosus

Gastrocnemius

Soleus

Achilles tendon

Levator scapulae

Rhomboideus minor

Rhomboideus major

Supraspinatus

Deltoideus

Infraspinatus

Teres major

Teres minor

Erector spinae

External intercostal

Internal abdominis oblique

Gluteus medius

Gluteus maximus

Adductor magnus

Gracilis

Biceps femoris

Figure 36b: Muscular system (posterior view).

Muscle Attachment

Skeletal (somatic or voluntary) muscles make up approximately 40% of the total human body weight. Their primary function is to produce movement through the ability to contract and relax in a co-ordinated manner. They are attached to bone by tendons. The place where a muscle attaches to a relatively stationary point on a bone, either directly or via a tendon, is called the *origin*. When the muscle contracts, it transmits tension to the bones across one or more joints, and movement occurs. The end of the muscle that attaches to the bone that moves is called the *insertion*.

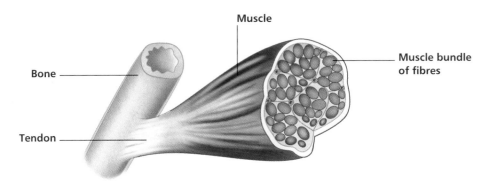

Figure 37: A tendon attachment.

Tendons and Aponeurosis

Muscle fascia, which is the connective tissue components of a muscle, combine together and extend beyond the end of the muscle as round cords or flat bands, called tendons; or as a thin, flat and broad aponeurosis. The tendon or aponeurosis secures the muscle to the bone or cartilage, to other muscles, or to a seam of fibrous tissue called a *raphe*.

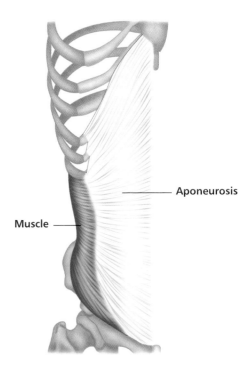

Figure 38: An attachment by aponeurosis.

Intermuscular Septa

In some cases, flat sheets of dense connective tissue known as intermuscular septa penetrate between muscles, providing another medium to which muscle fibres may attach.

Sesamoid Bones

If a tendon is subject to friction, it may, but not necessarily, develop a sesamoid bone within its substance. An example is the peroneus longus tendon in the sole of the foot. However, sesamoid bones may also appear in tendons not subject to friction.

Multiple Attachments

Many muscles have only two attachments, one at each end. However, more complex muscles are often attached to several different structures at its origin and/or its insertion. If these attachments are separated, effectively meaning the muscle gives rise to two or more tendons and/or aponeurosis inserting into different places, the muscle is said to have two heads. For example, the biceps brachii has two heads at its origin; one from the corocoid process of the scapula and one from the supraglenoid tubercle (*see* page 25). The triceps has three heads and the quadriceps has four.

Isometric and Isotonic Contractions

A muscle will contract upon stimulation, in an attempt to bring its attachments closer together, but this does not necessarily result in a shortening of the muscle. If the contraction of muscle results in the muscle creating movement of some sort, the contraction is called *isotonic*. If no movement results from contraction, such a contraction is called *isometric*.

Isometric

An isometric contraction occurs when a muscle increases its tension, but the length of the muscle is not altered. In other words, although the muscle tenses, the joint over which the muscle works does not move. One example of this is holding a heavy object in the hand with the elbow held stationary and bent at 90 degrees. Trying to lift something that proves to be too heavy to move is another example. Note also that some of the postural muscles are largely working isometrically by automatic reflex. For example, in the upright position, the body has a natural tendency to fall forward at the ankle. This is prevented by isometric contraction of the calf muscles. Likewise, the centre of gravity of the skull would make the head tilt forwards if the muscles at the back of the neck did not contract isometrically to keep the head centralized.

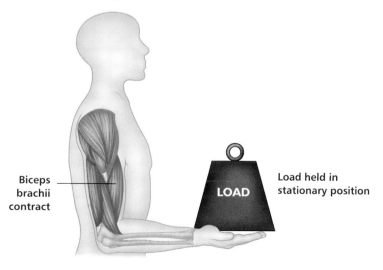

Figure 39: Isometric contraction.

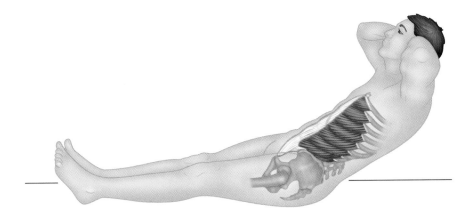

Figure 40: Abdominals contract to raise body concentrically.

Isotonic

It is the isotonic contractions of muscle that enable us to move about. Such contractions are of two types:

Concentric

In concentric contractions, the muscle attachments move closer together, causing movement at the joint. Using the example of holding an object in the hand, if the biceps brachii muscle contracts concentrically, the elbow joint will flex and the hand will move towards the shoulder, against gravity. Similarly, if we do a sit-up exercise, the abdominal muscles must contract concentrically to raise the torso (*see* figure 40).

Eccentric

Eccentric contraction means that the muscle fibres 'pay out' in a controlled manner to slow down movements which gravity, if unchecked, would otherwise cause to be too rapid. For example, lowering an object held in the hand down to your side. Another example is simply sitting down into a chair or lowering the torso after a sit-up exercise. Therefore, the difference between concentric and eccentric contraction is that in the former, the muscle shortens, and in the latter, it actually lengthens.

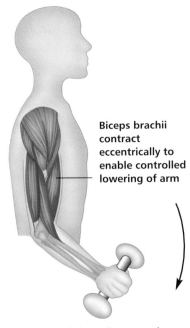

Biceps brachii contract eccentrically to enable controlled lowering of arm

Figure 41: Eccentric isotonic contraction.

Group Action of Muscles

Muscles work together, or in opposition, to achieve a wide variety of movements. Therefore, whatever one muscle can do, there is another muscle that can undo it. Muscles may also be required to provide additional support or stability to enable certain movements to occur elsewhere.

Muscles are classified into four functional groups:

1. Prime Mover or Agonist
2. Antagonist
3. Synergist
4. Fixator

Prime Mover or Agonist

A *prime mover* (*also* called an *agonist*) is a muscle that contracts to produce a specified movement. An example is the biceps brachii, which is the prime mover of elbow flexion. Other muscles may assist the prime mover in providing the same movement, albeit with less effect. Such muscles are called *assistant* or *secondary movers*. For example, the brachialis assists the biceps brachii in flexing the elbow, and is therefore a secondary mover.

Antagonist

The muscle on the opposite side of a joint to the prime mover, and which must relax to allow the prime mover to contract, is called an *antagonist*. For example, when the biceps brachii on the front of the arm contract to flex the elbow, the triceps brachii on the back of the arm must relax to allow this movement to occur. When the movement is reversed, i.e. when the elbow is extended, the triceps brachii becomes the prime mover and the biceps brachii assumes the role of antagonist.

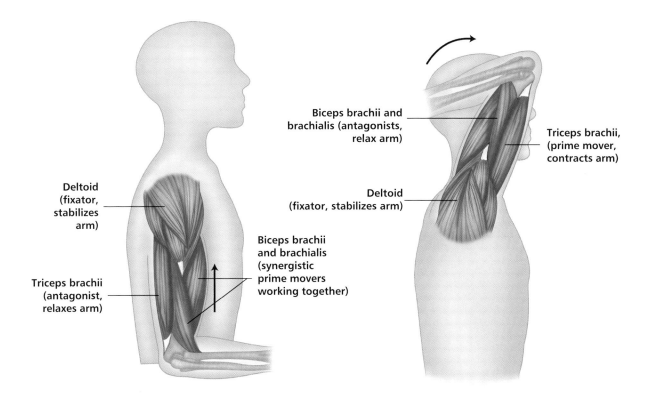

Figure 42: Group action of muscles;
a) flexing arm at elbow, and b) extending arm at elbow (showing reversed roles of prime mover and antagonist).

Synergist

Synergists prevent any unwanted movements that might occur as the prime mover contracts. This is especially important where a prime mover crosses two joints, because when it contracts it will cause movement at both joints, unless other muscles act to stabilize one of the joints. For example, the muscles that flex the fingers not only cross the finger joints, but also cross the wrist joint, potentially causing movement at both joints. However, it is because you have other muscles acting synergistically to stabilize the wrist joint that you are able to flex the fingers into a fist without also flexing the wrist at the same time.

A prime mover may have more than one action, so synergists also act to eliminate the unwanted movements. For example, the biceps brachii will flex the elbow, but its line of pull will also supinate the forearm (twist the forearm, as in tightening a screw). If you want flexion to occur without supination, other muscles must contract to prevent this supination. In this context, such synergists are sometimes called *neutralisers*.

Fixator

A synergist is more specifically referred to as a *fixator* or *stabilizer* when it immobilizes the bone of the prime mover's origin, thus providing a stable base for the action of the prime mover. The muscles that stabilize (fix) the scapula during movements of the upper limb are good examples. The sit-up exercise gives another good example. The abdominal muscles attach to both the ribcage and the pelvis. When they contract to enable you to perform a sit-up, the hip flexors will contract synergistically as fixators to prevent the abdominals tilting the pelvis; enabling the upper body to curl forward as the pelvis remains stationary.

2

Muscles of the Face, Head and Neck

EPICRANIUS (OCCIPITOFRONTALIS)

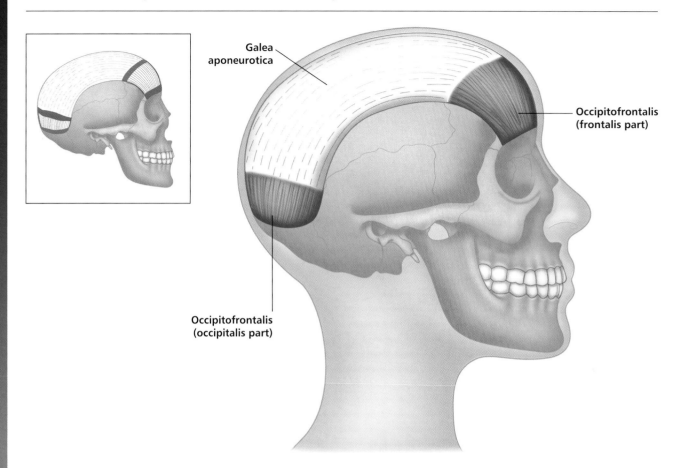

Galea aponeurotica

Occipitofrontalis (frontalis part)

Occipitofrontalis (occipitalis part)

Greek, *epi-*, above, upon; *cranium*, skull.

This muscle is effectively two muscles (occipitalis and frontalis), united by an aponeurosis called the **galea aponeurotica**, so named because it forms what resembles a helmet upon the skull.

Origin
Occipitalis: Occipital bone. Mastoid process of temporal bone.
Frontalis: Galea aponeurotica.

Insertion
Occipitalis: Galea aponeurotica (a sheet-like tendon leading to frontal belly).
Frontalis: Fascia and skin above eyes and nose.

Action
Occipitalis: Pulls scalp backward.
Frontalis: Pulls scalp forwards.

Nerve
Facial **V11** nerve.

Basic functional movement
Example: Raises eyebrows (wrinkles skin of forehead horizontally).

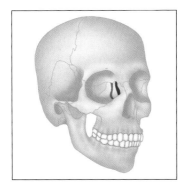

Lacrimal part

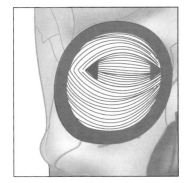

Orbital and palpebral part

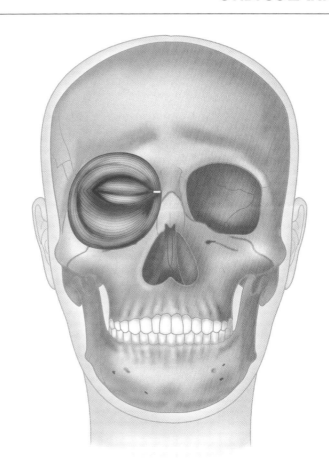

Latin, *orbis*, orb, circle; *oculi*, of the eye.

This complex and extremely important muscle consists of three parts, which together form an important protective mechanism surrounding the eye.

ORBITAL PART

Origin
Frontal bone. Medial wall of orbit (on maxilla).

Insertion
Circular path around orbit, returning to origin.

Action
Strongly closes eyelids (firmly 'screws up' the eye).

Nerve
Facial **V11** nerve (temporal and zygomatic branches).

PALPEBRAL PART
(in eyelids)

Latin, pertaining to an eyelid.

Origin
Medial palpebral ligament.

Insertion
Lateral palpebral ligament into zygomatic bone.

Action
Gently closes eyelids (and comes into action involuntarily, as in blinking).

Nerve
Facial **V11** nerve (temporal and zygomatic branches).

LACRIMAL PART
(behind medial palpebral ligament and lacrimal sac)

Latin, pertaining to the tears.

Origin
Lacrimal bone.

Insertion
Lateral palpebral raphe.

Action
Dilates lacrimal sac and brings lacrimal canals onto surface of eye.

Nerve
Facial **V11** nerve (temporal and zygomatic branches).

CORRUGATOR SUPERCILII

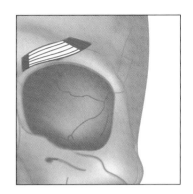

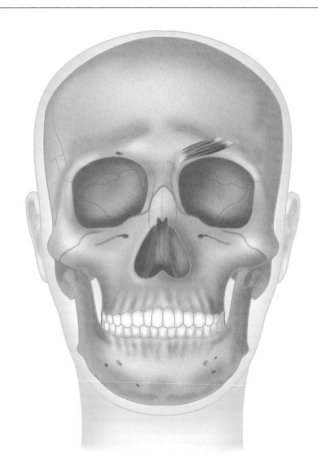

Latin, *corrugator*, muscle which wrinkles; *supercilii*, of the eyebrow.

Origin
Medial end of supercilliary arch of frontal bone.

Insertion
Deep surface of skin under medial half of the eyebrows.

Action
Draws eyebrows medially and downward, so producing vertical wrinkles, as in frowning.

Nerve
Facial **V11** nerve (temporal branch).

Basic functional movement
Facilitates facial expression.

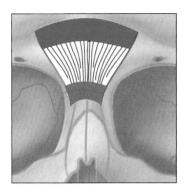

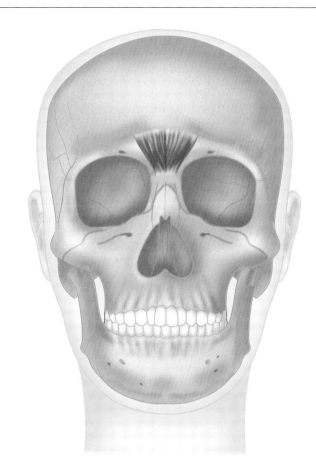

Latin, long, slender.

Origin
Fascia over nasal bone. Lateral nasal cartilage.

Insertion
Skin between eyebrows.

Action
Wrinkles nose. Pulls medial portion of eyebrows downwards.

Nerve
Facial **V11** nerve.

Basic functional movement
Example: Enables strong 'sniffing' and sneezing.

NASALIS

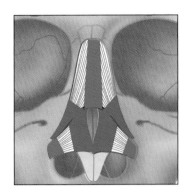

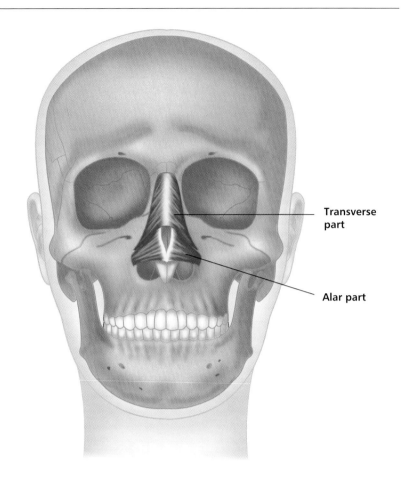

Transverse part

Alar part

Latin, *nasus*, nose.

Origin
Middle of maxilla (above incisor and canine teeth). Greater alar cartilage. Skin on nose.

Insertion
Joins muscle of opposite side across bridge of nose. Skin at tip of nose.

Action
Maintains opening of external nares during forceful inhalation (i.e. flares the nostrils).

Nerve
Facial **V11** nerve (buccal branches).

Basic functional movement
Example: Strongly breathing in through the nose.

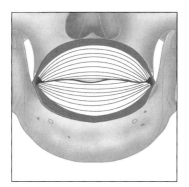

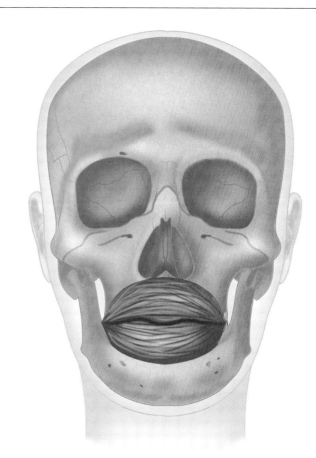

Latin, *orbis*, orb, circle; *oris*, pertaining to the mouth.

This is a composite sphincter muscle that encircles the mouth. It receives fasciculi from many other muscles.

Origin
Muscle fibres surrounding the opening of mouth, attached to the skin, muscle and fascia of the lips and surrounding area.

Insertion
Skin and fascia at corner of mouth.

Action
Closes lips, compresses lips against teeth, protrudes (purses) lips, and shapes lips during speech.

Nerve
Facial **V11** nerve (buccal and mandibular branches).

Basic functional movement
Facial expressions involving the lips.

LEVATOR LABII SUPERIORIS

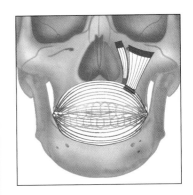

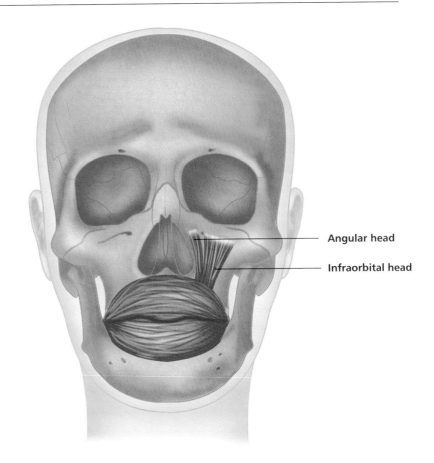

Angular head

Infraorbital head

Latin, *levare*, to raise; *labium*, lip; *superioris*, above.

Origin
Angular head: Zygomatic bone and frontal process of maxilla.
Infraorbital head: Lower border of orbit.

Insertion
Angular head: Greater alar cartilage, upper lip and skin of nose.
Infraorbital head: Muscles of upper lip.

Action
Raises upper lip. Dilates nares. Forms nasolabial furrow.

Nerve
Facial **V11** nerve (buccal branches).

Basic functional movement
Facilitates facial expression and kissing.

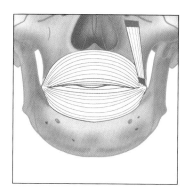

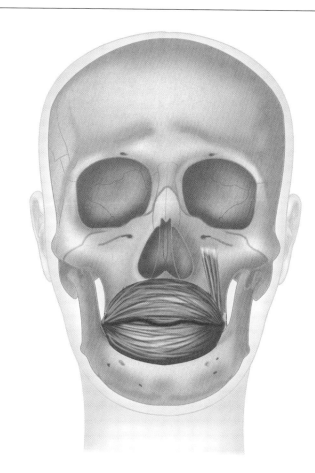

Latin, *levare*, to raise; *angulus*, angle; *oris*, pertaining to the mouth.

Origin
Canine fossa of maxilla.

Insertion
Corner of mouth.

Action
Elevates angle (corner) of mouth.

Nerve
Facial **V11** nerve (buccal branches).

Basic functional movement
Helps produce a smiling expression.

ZYGOMATICUS (MAJOR AND MINOR)

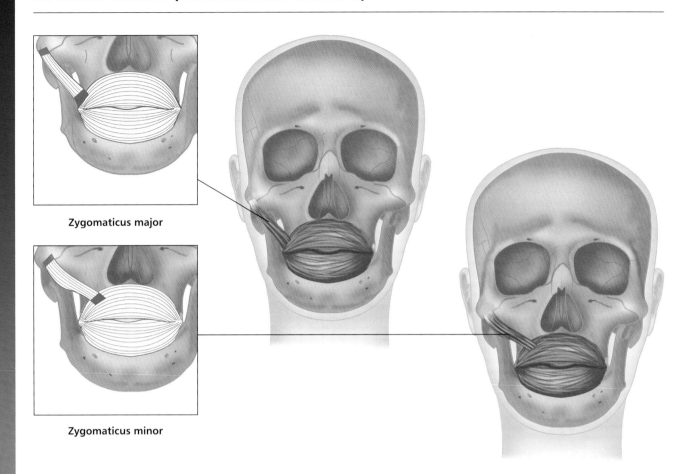

Zygomaticus major

Zygomaticus minor

Greek, *zygon*, yoke, union; **Latin**, *major*, large; *minor*, small.

Origin
Zygomaticus major: Upper lateral surface of zygomatic bone.
Zygomaticus minor: Lower surface of zygomatic bone.

Insertion
Zygomaticus major: Skin at corner of mouth. Orbicularis oris.
Zygomaticus minor: Lateral part of upper lip lateral to levator labii superioris.

Action
Zygomaticus major: Pulls corner of mouth up and back, as in smiling.
Zygomaticus minor: Elevates the upper lip. Forms nasolabial furrow.

Nerve
Facial **V11** nerve (zygomatic and buccal branches).

Basic functional movement
Smiling. Facilitates facial expression.

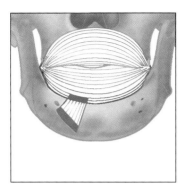

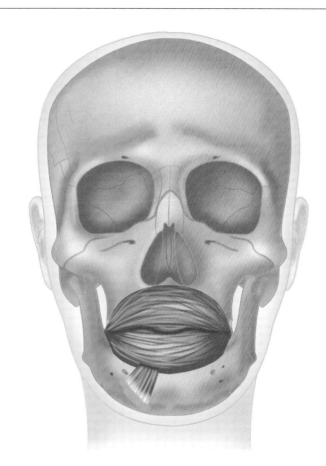

Latin, *deprimere*, to press down; *labii*, of the lip; *inferior*, below.

Origin
Anterior surface of mandible, between mental foramen and symphysis.

Insertion
Skin of lower lip.

Action
Pulls lower lip downward and slightly laterally.

Nerve
Facial **V11** nerve (marginal mandibular branch).

Basic functional movement
Facilitates facial expression.

DEPRESSOR ANGULI ORIS

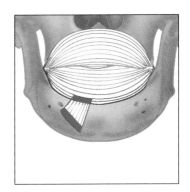

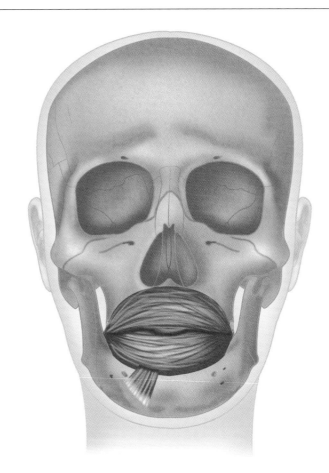

Latin, *deprimere*, to press down; *angulus*, angle; *oris*, pertaining to the mouth.

Muscle fibres are continuous with the platysma.

Origin
Oblique line of the mandible.

Insertion
Corner of mouth.

Action
Pulls corner of mouth downwards, as in sadness or frowning.

Nerve
Facial **V11** nerve (marginal mandibular and buccal branches).

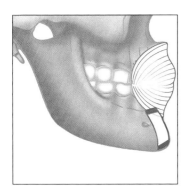

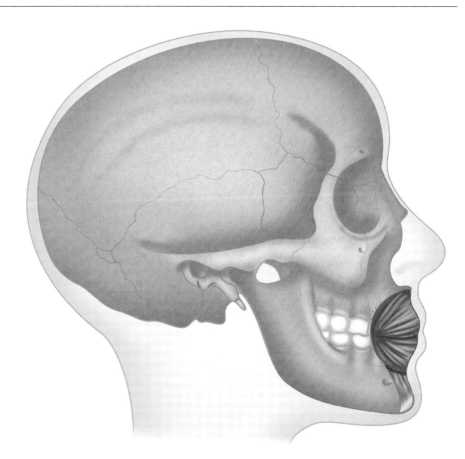

Latin, relating to the chin.

This is the only muscle of the lips that normally has no connection with the orbicularis oris.

Origin
Incisive fossa of anterior surface of mandible.

Insertion
Skin of chin.

Action
Protrudes lower lip and pulls up (wrinkles) skin of chin, as in pouting.

Nerve
Facial **V11** nerve (marginal mandibular branch).

PLATYSMA

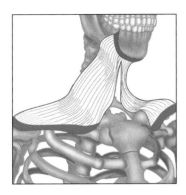

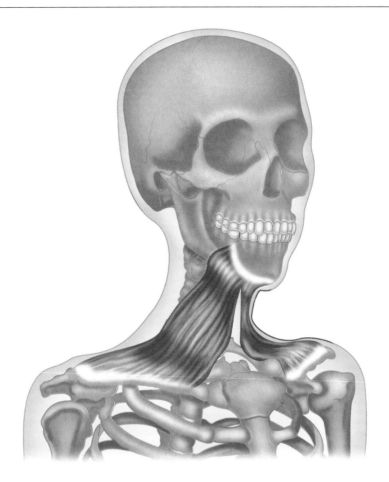

Greek, *platy*, broad, flat.

This muscle may be seen to stand out in a runner finishing a hard race.

Origin
Subcutaneous fascia of upper quarter of chest (i.e. fascia overlying the pectoralis major and deltoideus muscles).

Insertion
Subcutaneous fascia and muscles of chin and jaw. Inferior border of mandible.

Action
Pulls lower lip from corner of mouth downwards and laterally. Draws skin of chest upwards.

Nerve
Facial **V11** nerve (cervical branch).

Basic functional movement
Example: Gives expression of being startled or of sudden fright.

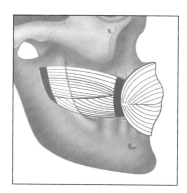

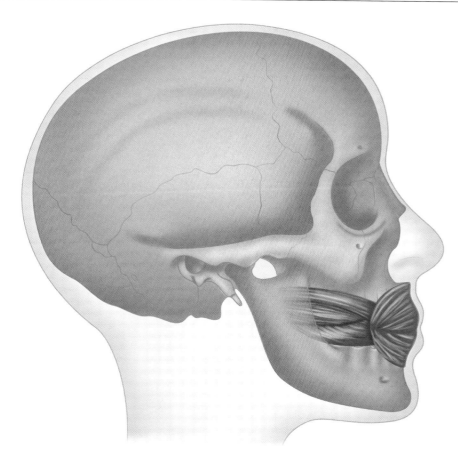

Latin, *buccina*, trumpet; *bucca*, cheek.

This muscle forms the substance of the cheek.

Origin
Alveolar processes of maxilla and mandible over molars and along pterygomandibular raphe (fibrous band extending from the pterygoid hamulus to the mandible).

Insertion
Orbicularis oris (muscles of lips).

Action
Compresses cheek as in blowing air out of mouth, and caves cheeks in, producing the action of sucking.

Nerve
Facial **V11** nerve (buccal branches).

MASSETER

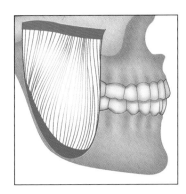

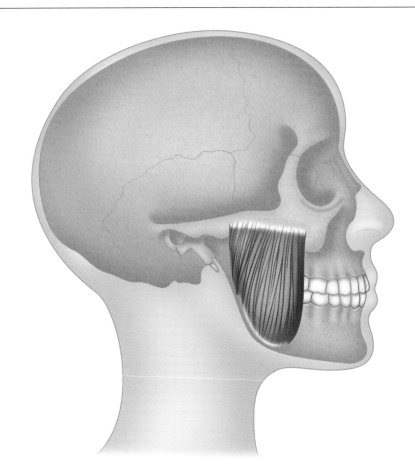

Greek, *maseter*, chewer.

The masseter is the most superficial muscle of mastication, easily felt when the jaw is clenched.

Origin
Zygomatic arch (cheek bone).

Insertion
Lateral surface of mandible (lower jaw).

Action
Closes jaw. Clenches teeth. Assists in side to side movement of mandible.

Nerve
Trigeminal **V** nerve (mandibular division).

Basic functional movement
Chewing food.

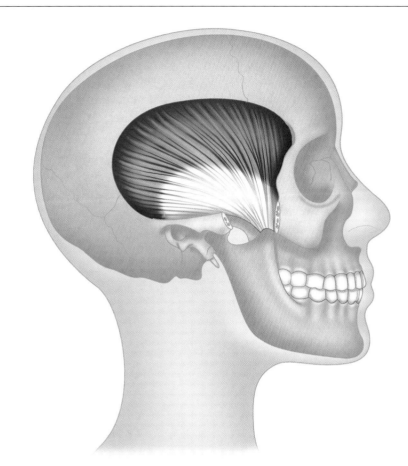

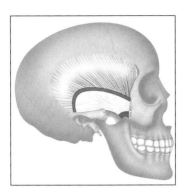

Zygomatic arch
has been removed.

Latin, time (seen by the greying of hairs in this region).

Origin
Temporal fossa including frontal, parietal and temporal bones.

Insertion
Coronoid process and ramus of mandible (area on lower jaw, just below the lateral edge of the zygomatic arch).

Action
Closes jaw. Clenches teeth. Assists in side to side movement of mandible.

Nerve
Anterior and posterior deep temporal nerves from the trigeminal **V** nerve (mandibular division).

Basic functional movement
Chewing food.

PTERYGOIDEUS LATERALIS (LATERAL PTERYGOID)

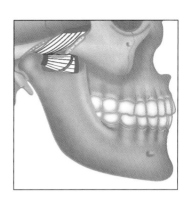

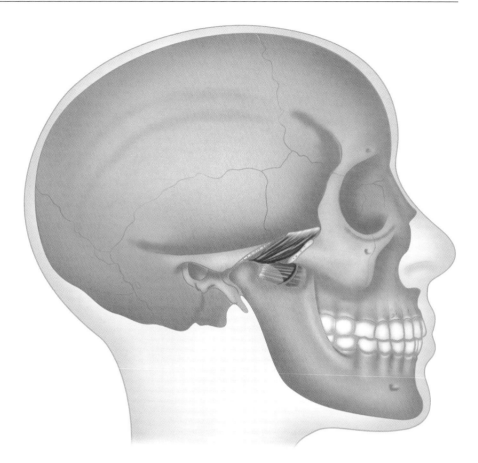

Greek, *pterygodes*, like a wing; **Latin**, *lateral*, to the side.

The superior head of this muscle is sometimes called *sphenomeniscus*, because it inserts into the disc of the temporomandibular joint.

Origin
Superior head: Lateral surface of greater wing of sphenoid.
Inferior head: Lateral surface of lateral pterygoid plate of sphenoid.

Insertion
Superior head: Capsule and articular disc of the temporomandibular joint.
Inferior head: Neck of mandible.

Action
Protrudes mandible. Opens mouth. Moves mandible from side to side (as in chewing).

Nerve
Trigeminal **V** nerve (mandibular division).

Basic functional movement
Chewing food.

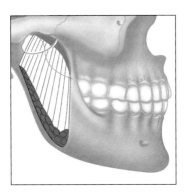

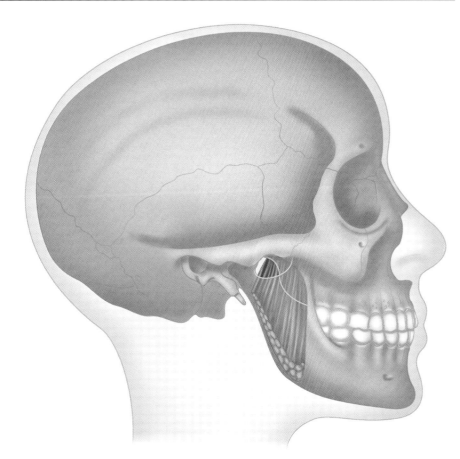

Greek, *pterygodes*, like a wing; **Latin**, *medius*, middle.

This muscle mirrors the masseter muscle in both its position and action, with the ramus of the mandible positioned between the two muscles.

Origin
Medial surface of lateral pterygoid plate of the sphenoid bone. Pyramidal process of the palatine bone. Tuberosity of maxilla.

Insertion
Medial surface of the ramus and the angle of the mandible.

Action
Elevates and protrudes the mandible. Therefore it closes the jaw and assists in side to side movement of the mandible, as in chewing.

Nerve
Trigeminal **V** nerve (mandibular division).

Basic functional movement
Chewing food.

SCALENUS ANTERIOR, MEDIUS, POSTERIOR

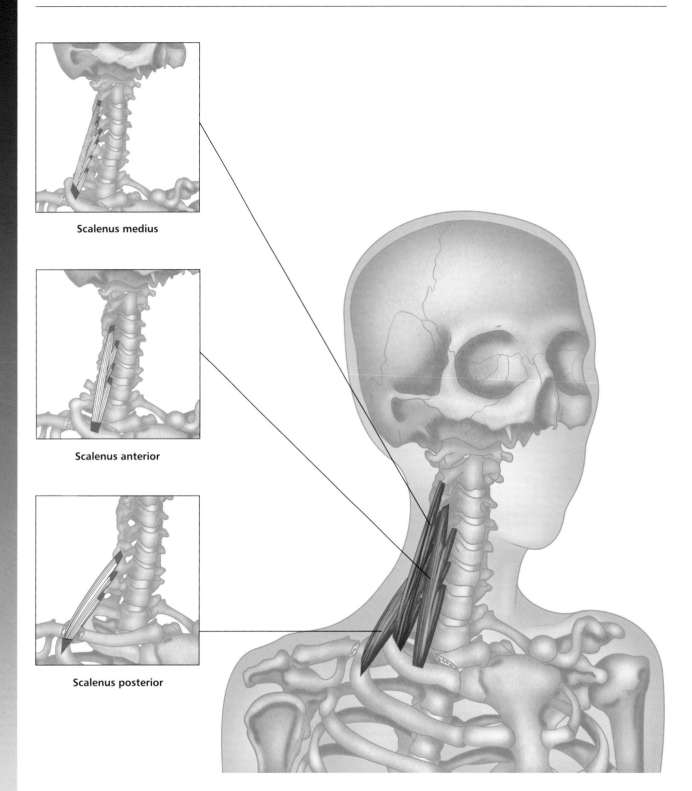

Scalenus medius

Scalenus anterior

Scalenus posterior

Strengthening exercise

Twisting sit-ups

Greek, *skalenos*, uneven; **Latin**, *anterior*, before; *medius*, middle; *posterior*, behind.

Origin
Transverse processes of cervical vertebrae.

Insertion
Anterior and medius: First rib.
Posterior: Second rib.

Action
Acting together: Flex neck. Raise first rib during a strong inhalation.
Individually: Laterally flex and rotate neck.

Nerve
Ventral rami of cervical nerves, C3–C8.

Basic functional movement
The scaleni are primarily muscles of inspiration.

Sports that heavily utilise these muscles
All active sports that require strong respiration (e.g. high pace running).

Common problems when muscles are chronically tight / shortened
Painful conditions of the neck, shoulder and arm because hypertonic muscle puts pressure on a bundle of nerves called the *brachial plexus,* and the subclavian artery.

Self stretches

Pull left shoulder
away from ear. Do not
rotate head. Draw right ear
towards right shoulder.

STERNOCLEIDOMASTOIDEUS

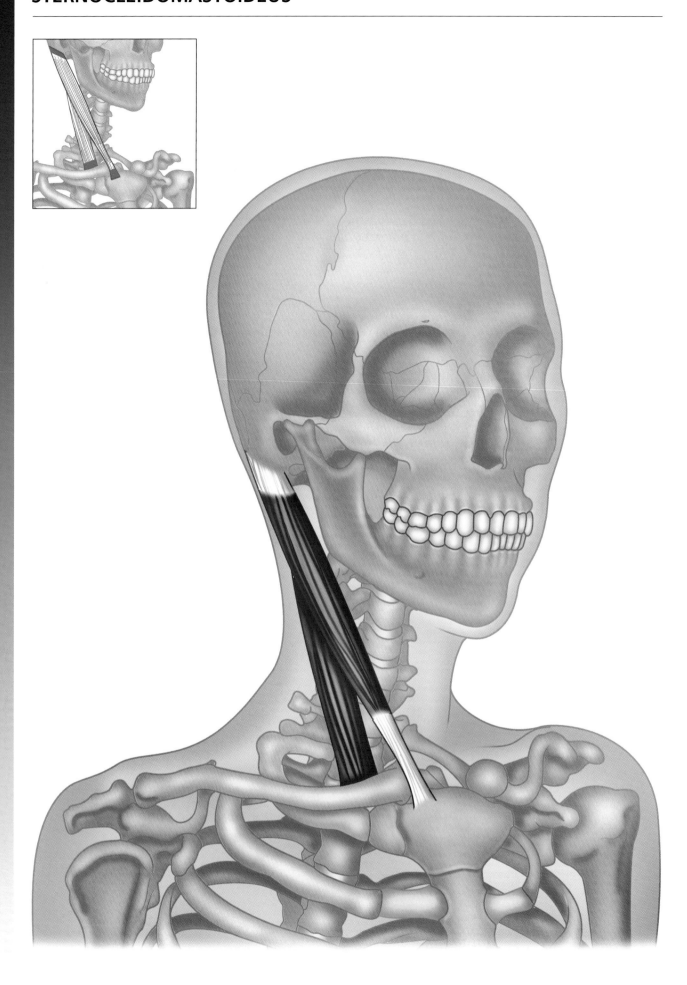

Strengthening exercise

Sit-ups

Greek, *sternon*, sternum; *kleidos*, key, clavicle; *mastoid*, breast-shaped, mastoid process.

This muscle is a long strap muscle with two heads. It is sometimes injured at birth, and may be partly replaced by fibrous tissue that contracts to produce a torticollis (wry neck).

Origin
Sternal head: Anterior surface of upper sternum.
Clavicular head: Medial third of clavicle.

Insertion
Mastoid process of temporal bone (bony prominence just behind the ear).

Action
Contraction of both sides together: Flexes neck (draws head forward). Raises sternum, and consequently the ribs, during deep inhalation.
Contraction of one side: Tilts the head towards the same side. Rotates head to face the opposite side (and also upward as it does so).

Nerve
Accessory **X1** nerve; with sensory supply for proprioception from cervical nerves C2 and C3.

Basic functional movement
Examples: Turning head to look over your shoulder. Raising head from pillow.

Sports that heavily utilise this muscle
Examples: Swimming. Rugby scrummage. American football.

Movements or injuries that may damage this muscle
Extreme whiplash movements.

Common problems when muscle is chronically tight / shortened
Headache and neck pain.

Self stretch

Turn head to right.
Repeat on opposite side.

3

Muscles of
the Trunk

ERECTOR SPINAE (SACROSPINALIS)

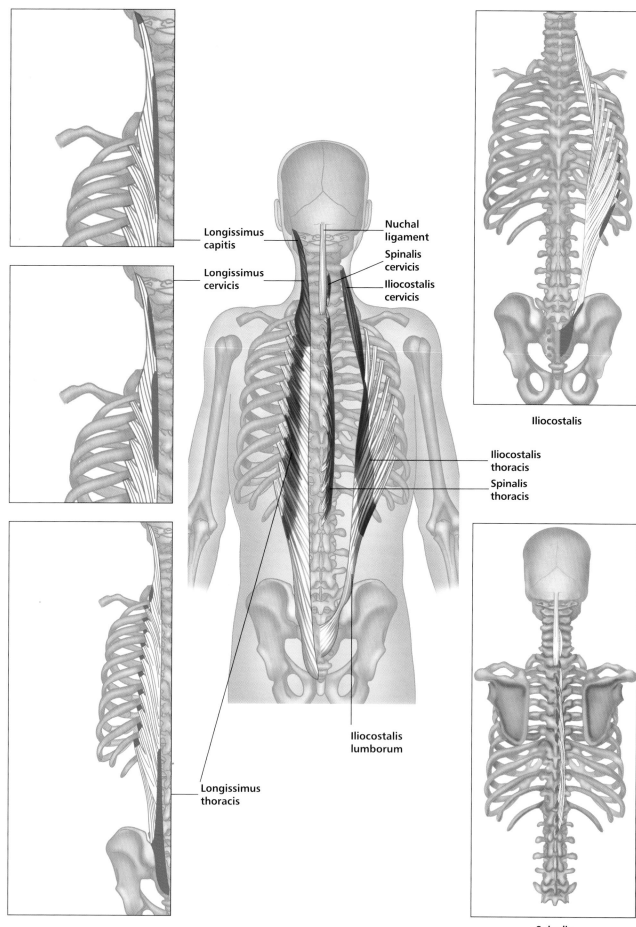

Longissimus
capitis

Longissimus
cervicis

Nuchal
ligament

Spinalis
cervicis

Iliocostalis
cervicis

Iliocostalis
thoracis

Spinalis
thoracis

Iliocostalis
lumborum

Longissimus
thoracis

Iliocostalis

Spinalis

Strengthening exercises

Back extension (back raise)

Lat. pull-downs

Squats

Squats

Side bends

Latin, *sacrum*, sacred; *spinalis*, spinal.

The erector spinae, also called *sacrospinalis*, comprises three sets of muscles organised in parallel columns. From lateral to medial, they are: iliocostalis, longissimus and spinalis.

Origin
Slips of muscle arising from the sacrum. Iliac crest. Spinous and transverse processes of vertebrae. Ribs.

Insertion
Ribs. Transverse and spinous processes of vertebrae. Occipital bone.

Action
Extends and laterally flexes vertebral column (i.e. bending backwards and sideways).
Helps maintain correct curvature of spine in the erect and sitting positions. Steadies the vertebral column on the pelvis during walking.

Nerve
Dorsal rami of cervical, thoracic and lumbar spinal nerves.

Basic functional movement
Keeps back straight (with correct curvatures). Therefore maintains posture.

Sports that heavily utilise these muscles
Examples: All sports, especially swimming, gymnastics, and wrestling.

Movements or injuries that may damage these muscles
Lifting without bending the knees or keeping the back erect, or holding the object too far in front of the body.

Self stretches

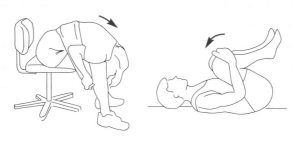

Move towel up back with each set of stretching.

Pull knees into your chest and up towards your shoulders.

SEMISPINALIS CAPITIS, CERVICIS, THORACIS

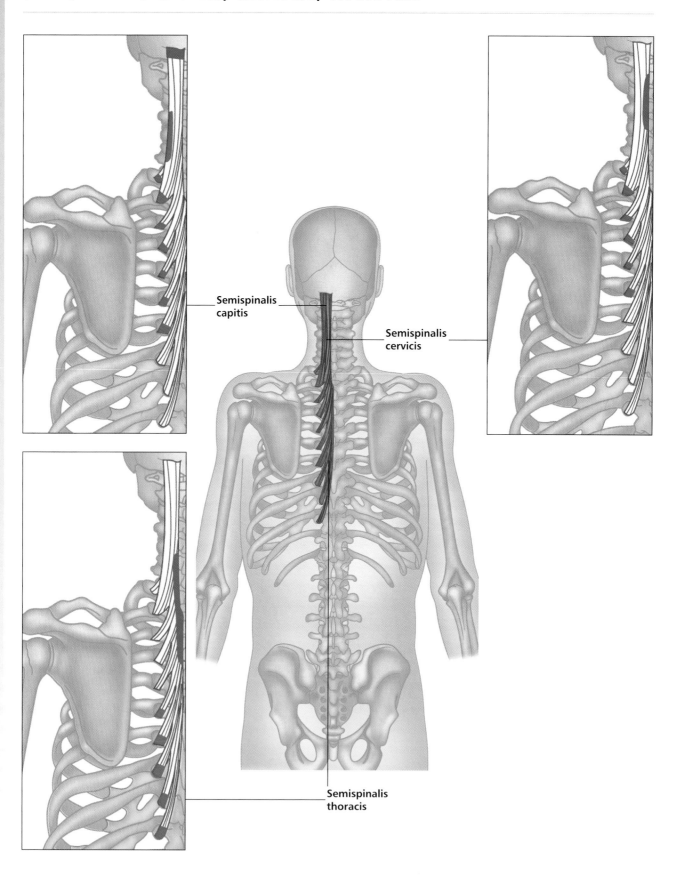

Semispinalis
capitis

Semispinalis
cervicis

Semispinalis
thoracis

Strengthening exercise

Back extension (back raise)

Latin, *semispinalis*, half spinal; *capitis*, of the head; *cervix*, neck; **Greek**, *thoracis*, chest.

The transversospinalis is a composite of three small muscle groups situated deep to erector spinae. However, unlike erector spinae, each group lies successively deeper from the surface rather than side-by-side. The muscle groups are, from more superficial to deep: semispinalis, multifidis, and rotatores. Their fibres generally extend upward and medially from transverse processes to higher spinous processes.

Origin
Transverse processes of cervical and thoracic vertebrae, (C1–T10).

Insertion
Between nuchal lines of occipital bone and spinous processes of the cervical vertebrae and upper four thoracic vertebrae, (C2–T4).

Action
Capitis: Most powerful extensor of the head and assists in rotation.
Cervicis and thoracis: Extends thoracic and cervical parts of vertebral column. Assists rotation of thoracic and cervical vertebrae.

Nerve
Dorsal rami of cervical and thoracic spinal nerves.

Basic functional movement
Looking up, or turning the head to look behind.

Sports that heavily utilise these muscles
Examples: Rugby scrums. American football. Wrestling. Swimming.

Movements or injuries that may damage these muscles
Whiplash injuries.

Self stretch

Arch your back as if being
drawn up by a piece of string.

SPLENIUS CAPITIS AND SPLENIUS CERVICIS

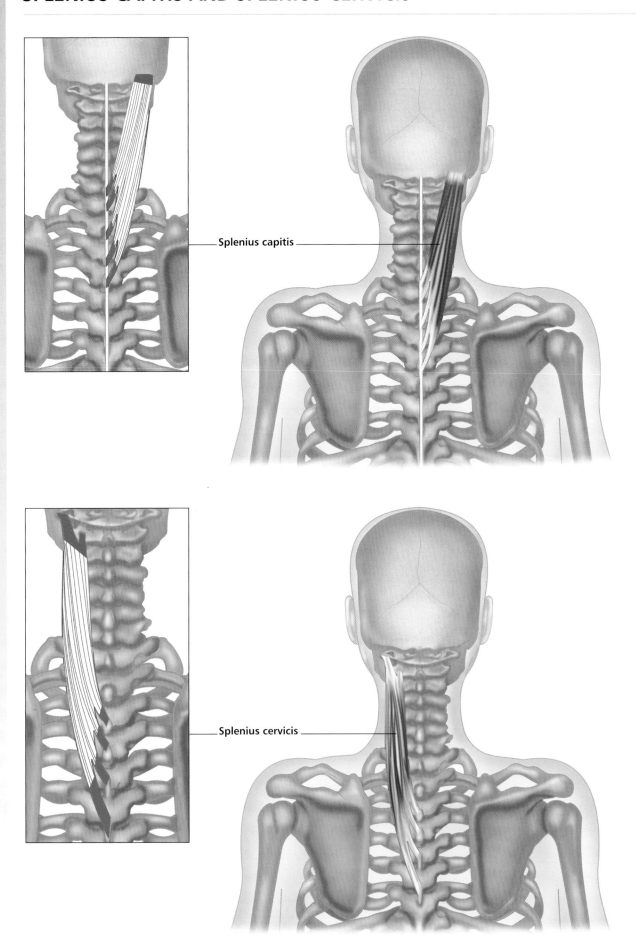

Splenius capitis

Splenius cervicis

Strengthening exercise

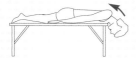

Greek, *splenion*, bandage; **Latin**, *capitis*, of the head; *cervix*, neck.

Origin
Splenius capitis: Lower part of ligamentum nuchae. Spinous processes of the seventh cervical vertebra, (C7) and upper three or four thoracic vertebrae, (T1–T4).
Splenius cervicis: Spinous processes of the third to sixth thoracic vertebrae, (T3–T6).

Insertion
Splenius capitis: Posterior aspect of mastoid process of temporal bone. Lateral part of superior nuchal line, deep to the attachment of the sternocleidomastoideus.
Splenius cervicis: Posterior tubercles of transverse processes of the upper two or three cervical vertebrae, (C1–C3).

Action
Acting together: Extend the head and neck.
Individually: Laterally flexes neck. Rotates the face to the same side as contracting muscle.

Nerve
Dorsal rami of middle and lower cervical nerves.

Basic functional movement
Example: Looking up, or turning the head to look behind.

Sports that heavily utilise these muscles
Examples: Rugby scrums. American football. Wrestling. Swimming.

Movements or injuries that may damage these muscles
Whiplash injuries.

Common problems when muscle is chronically tight / shortened
Headache and neck pain.

Self stretch

MULTIFIDIS

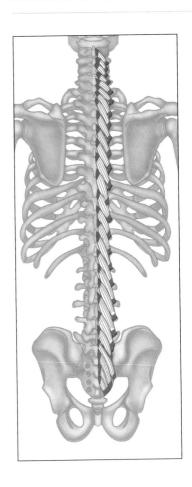

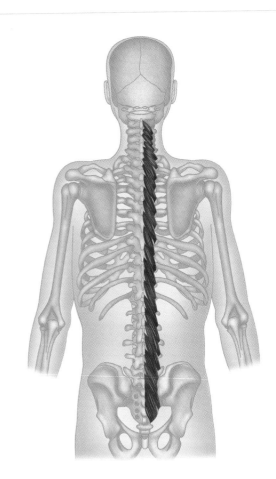

Latin, *multi*, many, much; *findere*, to split.

This muscle is the part of the transversospinalis group that lies in the furrow between the spines of the vertebrae and their transverse processes.

Origin
Posterior surface of sacrum, between the sacral foramina and posterior superior iliac spine. Mamillary processes (posterior borders of superior articular processes) of all lumbar vertebrae. Transverse processes of all thoracic vertebrae. Articular processes of lower four cervical vertebrae.

Insertion
Parts insert into spinous process two to four vertebrae superior to origin; overall including spinous processes of all the vertebrae from the fifth lumbar up to the axis, (L5–C2).

Action
Protects vertebral joints from movements made by the more powerful superficial prime movers. Extension, lateral flexion and rotation of vertebral column.

Nerve
Dorsal rami of spinal nerves.

Basic functional movement
Helps maintain good posture and spinal stability during all movements.

Movements or injuries that may damage this muscle
Lifting without bending the knees or keeping the back erect, or holding the object too far in front of the body.

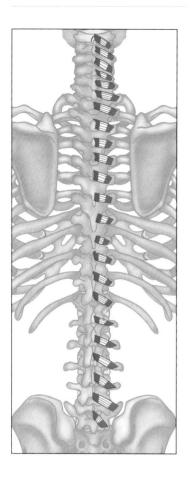

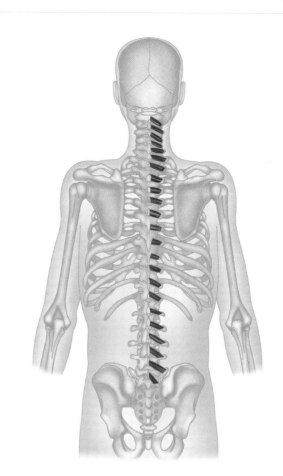

Latin, *rot*, wheel.

These small muscles are the deepest layer of the transversospinalis group.

Origin
Transverse process of each vertebra.

Insertion
Base of spinous process of adjoining vertebra above.

Action
Rotate and assist in extension of vertebral column.

Nerve
Dorsal rami of spinal nerves.

Basic functional movement
Helps maintain good posture and spinal stability during standing, sitting and all movements.

Movements or injuries that may damage this muscle
Lifting without bending the knees or keeping the back erect, or holding the object too far in front of the body.

EXTERNAL AND INTERNAL INTERCOSTALS

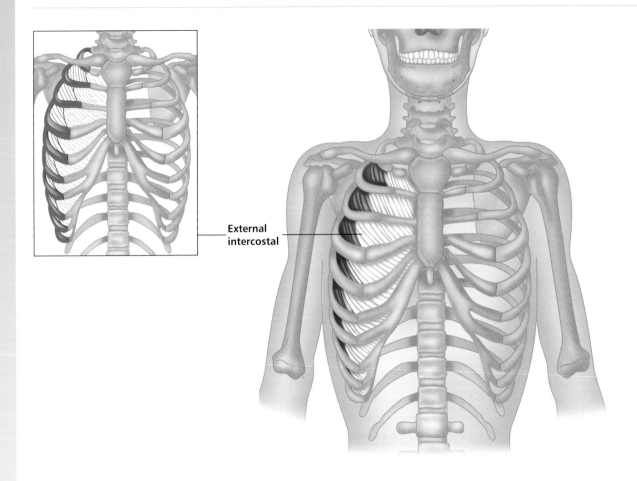

External
intercostal

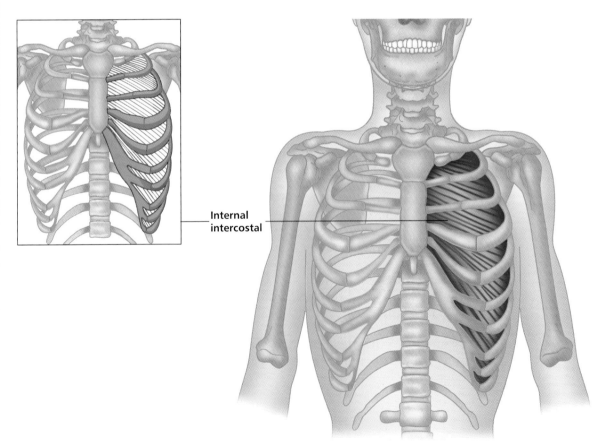

Internal
intercostal

Strengthening exercise

Twisting sit-ups

Latin, *inter*, between; *costal*, rib.

The lower external intercostal muscles may blend with the fibres of external oblique, which overlap them, thus effectively forming one continuous sheet of muscle, with the external intercostal fibres seemingly stranded between the ribs. There are 11 external intercostals on each side of the ribcage.

Internal intercostal fibres lie deep to, and run obliquely across, the external intercostals. There are 11 internal intercostals on each side of the ribcage.

Origin
External intercostals: Lower border of a rib.
Internal intercostals: Upper border of a rib and costal cartilage.

Insertion
External intercostals: Upper border of rib below (fibres runs obliquely forwards and downwards).
Internal intercostals: Lower border of rib above (fibres runs obliquely forwards and upwards towards the costal cartilage).

Action
Muscles contract to stabilize the ribcage during various movements of the trunk. Prevents the intercostal space from bulging out or sucking in during respiration.

Nerve
The corresponding intercostal nerves.

Sports that heavily utilise these muscles
All very active sports.

Common problems when muscles are chronically tight / shortened
Kyphosis (rounded back) and depressed chest.

Self stretch

Avoid or take care
if you have back problems;
check with your health
professional first.

DIAPHRAGM

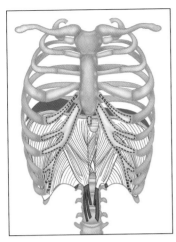

Origin on posterior of costal cartilage.

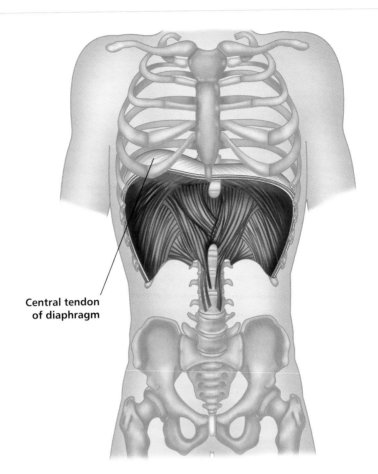

Central tendon of diaphragm

Greek, partition, wall.

Origin
Back of xiphoid process (lower tip of breastbone / sternum).
Lower six ribs and their costal cartilages.
Upper two or three lumbar vertebrae, (L1–L3).

Insertion
All fibres converge and attach onto a central tendon, i.e. this muscle inserts upon itself.

Action
Forms floor of thoracic cavity. Pulls its central tendon downward during inhalation, thereby increasing volume of thoracic cavity.

Nerve
Phrenic nerve (ventral rami), C3, **4**, 5.

Basic functional movement
Produces about 60% of your breathing capacity.

Sports that heavily utilise this muscle
All physically demanding sports.

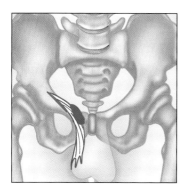

Anterior view.

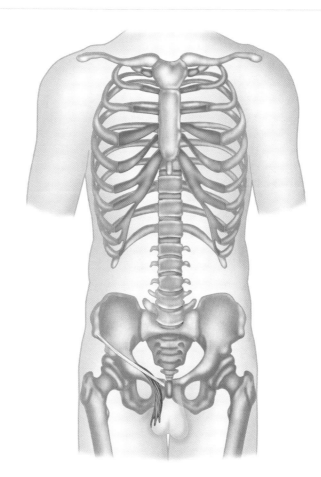

Greek, *kremasthai*, to suspend.

In males, the cremaster is usually well-developed. In females, it is underdeveloped or absent. It forms a thin network of muscle fibres around the spermatic cord and testes (or around the distal portion of the round ligament of the uterus).

Origin
Inguinal ligament.

Insertion
Pubic tubercle. Crest of pubis. Sheath of rectus abdominis.

Action
Pulls testes up from the scrotum towards the body (mainly to regulate the temperature of the testes).

Nerve
Genital branch of genitofemoral nerve, L1, 2.

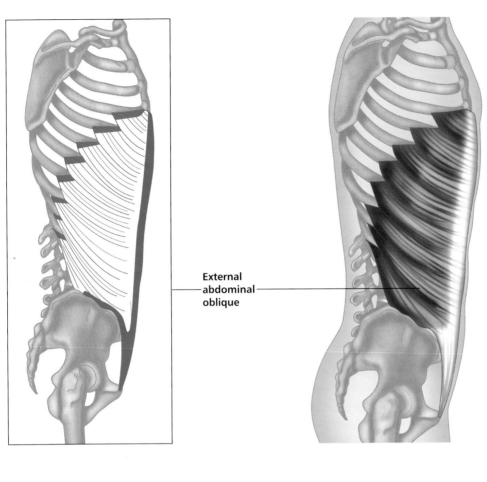

External
abdominal
oblique

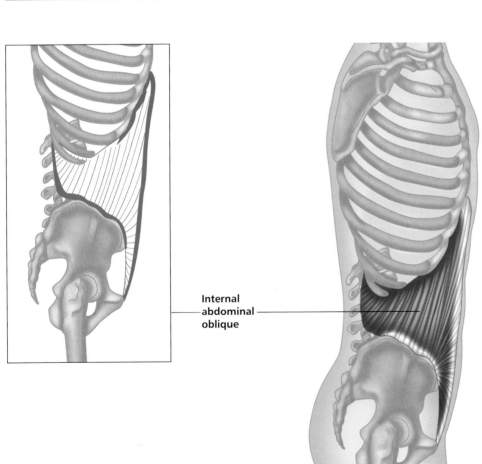

Internal
abdominal
oblique

Strengthening exercises

Twisting sit-ups

Abdominal machine crunch
(for upper fibres)

Hanging leg raise

Reverse trunk twist

Side bends

Latin, *obliquus*, diagonal, slanted.

The posterior fibres of the external oblique are usually overlapped by the latissimus dorsi, but in some cases there is a space between the two, known as the **lumbar triangle**, situated just above the iliac crest. The lumbar triangle is a weak point in the abdominal wall.

Origin
External oblique: Lower eight ribs.
Internal oblique: Iliac crest. Lateral two-thirds of inguinal ligament. Thoracolumbar fascia (i.e. sheet of connective tissue in lower back).

Insertion
External oblique: Anterior half of iliac crest, and into an abdominal aponeurosis that terminates in the linea alba (a tendinous band extending downwards from the sternum).
Internal oblique: Bottom three or four ribs, and linea alba via aponeurosis.

Action
Compresses abdomen, helping to support the abdominal viscera against the pull of gravity. Contraction of one side alone bends the trunk laterally to that side and rotates it to the opposite side.

Nerve
External oblique: Ventral rami of thoracic nerves, T5–T12.
Internal oblique: Ventral rami of thoracic nerves, T7–T12, ilioinguinal and iliohypogastric nerves.

Basic functional movement
Example: Digging with a shovel, raking.

Sports that heavily utilise these muscles
External obliques: Examples: Gymnastics. Rowing. Rugby.
Internal obliques: Examples: Golf. Javelin. Pole vault.

Common problems when muscles are weak
Injury to lumbar spine, because abdominal muscle tone contributes to stability of lumbar spine.

Self stretches

Try to twist using trunk rather than shoulders or arms.

Perform this exercise slowly, thus avoiding the tendency to use momentum.

Avoid or take care if you have back problems; check with your health professional first.

TRANSVERSUS ABDOMINIS

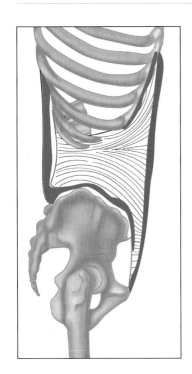

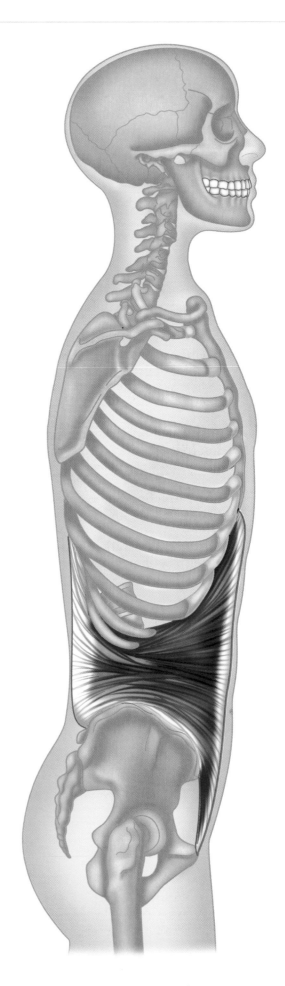

**Strengthening
exercise**

Twisting sit-ups

Latin, *transversus*, across, crosswise; *abdominis*, belly / stomach.

Origin
Anterior two thirds of iliac crest. Lateral third of inguinal ligament. Costal cartilages of lower six ribs. Thoracolumbar fascia.

Insertion
Linea alba via an abdominal aponeurosis (tendinous band extending between the sternum and pubis).

Action
Compresses abdomen, helping to support the abdominal viscera against the pull of gravity.

Nerve
Ventral rami of thoracic nerves T7–T12, ilioinguinal and iliohypogastric nerves.

Basic functional movement
Important during forced expiration, sneezing and coughing. Helps maintain good posture.

Sports that heavily utilise this muscle
Examples: Gymnastics. Seated rowing. Javelin. Pole vault.

Common problems when muscle is weak
Injury to lumbar spine, because abdominal muscle tone contributes to stability of lumbar spine.

Self stretches

Avoid or take care
if you have back problems;
check with your health
professional first.
However, this muscle is
rarely too tight.

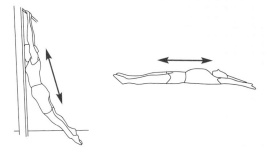

RECTUS ABDOMINIS

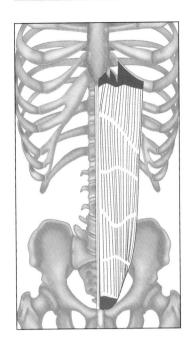

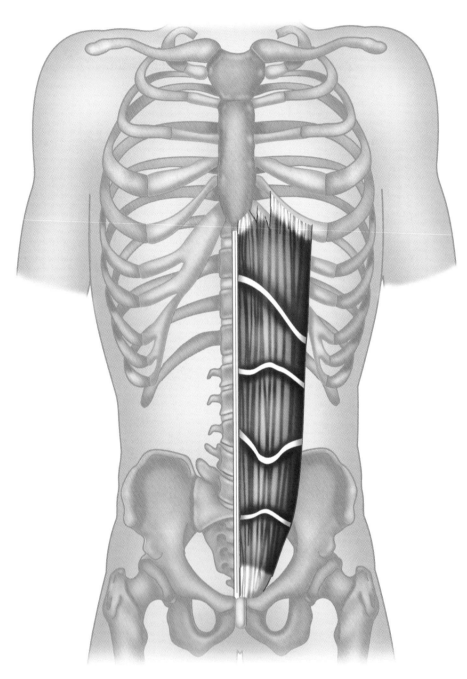

Strengthening exercises

Sit-ups

Abdominal machine crunch
(for upper fibres)

Reverse sit-up
(for lower fibres)

Hanging leg raise

Latin, *rectum*, straight; *abdominis*, belly / stomach.

The rectus abdominis is divided into three or four bellies by tendinous bands, each sheathed in aponeurotic fibres from the lateral abdominal muscles. These fibres converge centrally to form the linea alba. Situated anterior to the lower part of rectus abdominis is a frequently absent muscle called *pyramidalis*, which arises from the pubic crest and inserts into the linea alba. It tenses the linea alba, for reasons unknown. Associated with the *six-pack* muscles seen in conditioned athletes.

Origin
Pubic crest and symphysis (front of pubic bone).

Insertion
Xiphoid process (base of sternum). Fifth, sixth and seventh costal cartilages.

Action
Flexes lumbar spine. Depresses ribcage. Stabilizes the pelvis during walking.

Nerve
Ventral rami of thoracic nerves, T5–T12.

Basic functional movement
Example: Initiating getting out of a low chair.

Sports that heavily utilise this muscle
All sports.

Common problems when muscle is weak
Injury to lumbar spine, because abdominal muscle tone contributes to stability of lumbar spine.

Self stretches

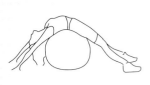

Avoid or take care
if you have back problems;
check with your health
professional first.

QUADRATUS LUMBORUM

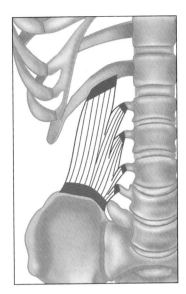

Anterior view.

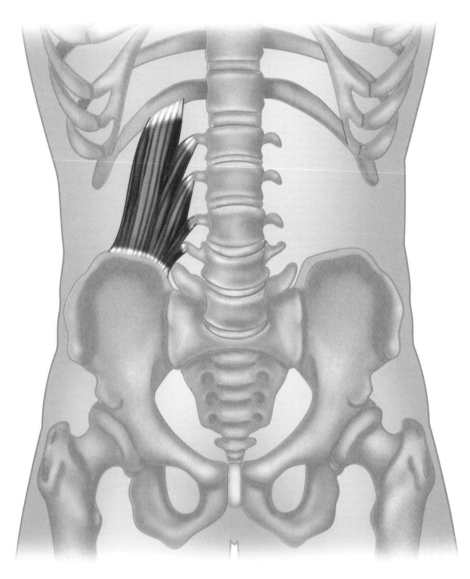

**Strengthening
exercise**

Side bends

Latin, *quadratus*, four-sided; *lumbar*, loin.

Origin
Iliac crest. Iliolumbar ligament (the ligament from the 5th lumbar vertebra to the ilium).

Insertion
Twelfth rib. Transverse processes of upper four lumbar vertebrae, (L1–L4).

Action
Laterally flexes vertebral column. Fixes the 12th rib during deep respiration (e.g. helps stabilize the diaphragm for singers exercising voice control). Helps extend lumbar part of vertebral column, and gives it lateral stability.

Nerve
Ventral rami of the subcostal nerve and upper three or four lumbar nerves, T12, L**1**, **2**, **3**.

Basic functional movement
Bending sideways from sitting to pick up an object from the floor.

Sports that heavily utilise this muscle.
Examples: Gymnastics (pommel horse). Javelin. Tennis serve.

Movements or injuries that may damage this muscle
Bending sideways or lifting from sideways position too quickly.

Common problems when muscle is chronically tight / shortened
Referred pain to hip and gluteal area, low back.

Self stretches

Place towel under
left foot. Side bend to left,
progressively taking up
any slack in towel.

ILIOPSOAS (PSOAS MAJOR AND ILIACUS)

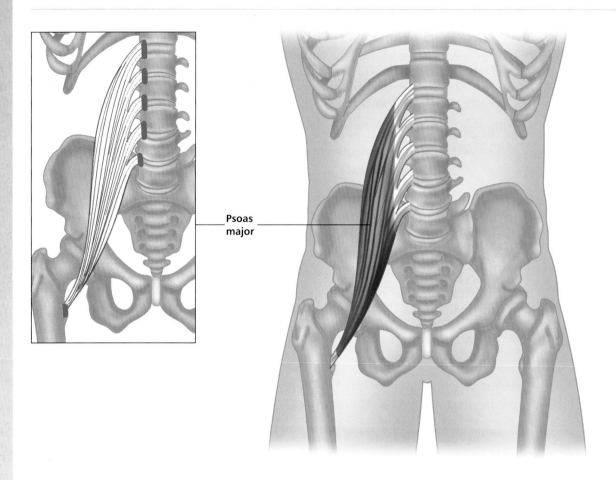

Psoas
major

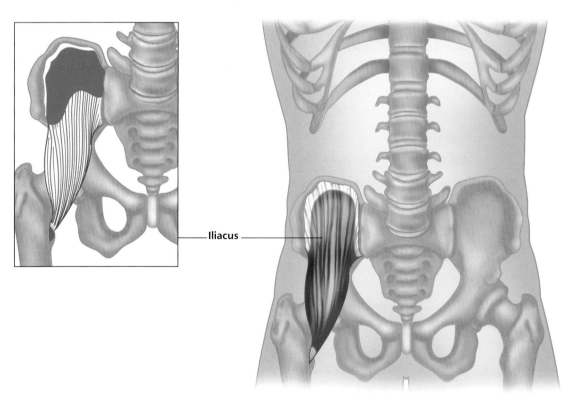

Iliacus

Strengthening exercises

Incline sit-ups

Hanging leg raise

Multi-hip machine
(hip joint flexion)

Greek, *psoas*, muscle of the loin; *major*, large; *iliacus*, pertaining to the loin.

The psoas major and iliacus are considered part of the posterior abdominal wall due to their position and cushioning role for the abdominal viscera. However, based on their action of flexing the hip joint, it would also be relevant to place them with the hip muscles (*see* page 143). Note that some upper fibres of psoas major may insert by a long tendon into the iliopubic eminence to form the psoas minor, which has little function and is absent in about 40% of people.

Bilateral contracture of this muscle will increase lumbar lordosis.

Origin
Psoas major: Transverse processes of all lumbar vertebrae, (L1–L5). Bodies of twelfth thoracic and all lumbar vertebrae, (T12–L5). Intervertebral discs above each lumbar vertebra.
Iliacus: Superior two-thirds of iliac fossa. Anterior ligaments of the lumbosacral and sacroiliac joints.

Insertion
Lesser trochanter of femur.

Action
Main flexor of hip joint (flexes and laterally rotates thigh, as in kicking a football). Acting from its insertion, flexes the trunk, as in sitting up from the supine position.

Nerve
Psoas major: Ventral rami of lumbar nerves, L1, **2**, **3**, 4.
Iliacus: Femoral nerve, L(1), **2**, **3**, 4.

Basic functional movement
Example: Going up a step or walking up an incline.

Sports that heavily utilise these muscles
Examples: Rock-face climbing. Sprinting (maximizes stride length). Kicking sports (e.g. soccer, to maximise kicking force).

Common problems when muscles are chronically tight / shortened
Low back pain due to increase in lumbar curve (lordosis).

Self stretch

Push right hip forward
to stretch right iliopsoas.
Keep low back flat and
maintain upright posture.

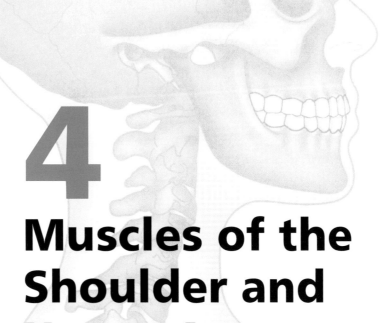

4

Muscles of the Shoulder and Upper Arm

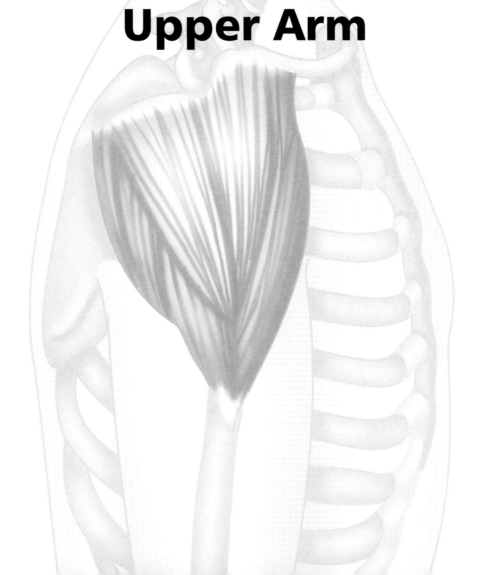

TRAPEZIUS

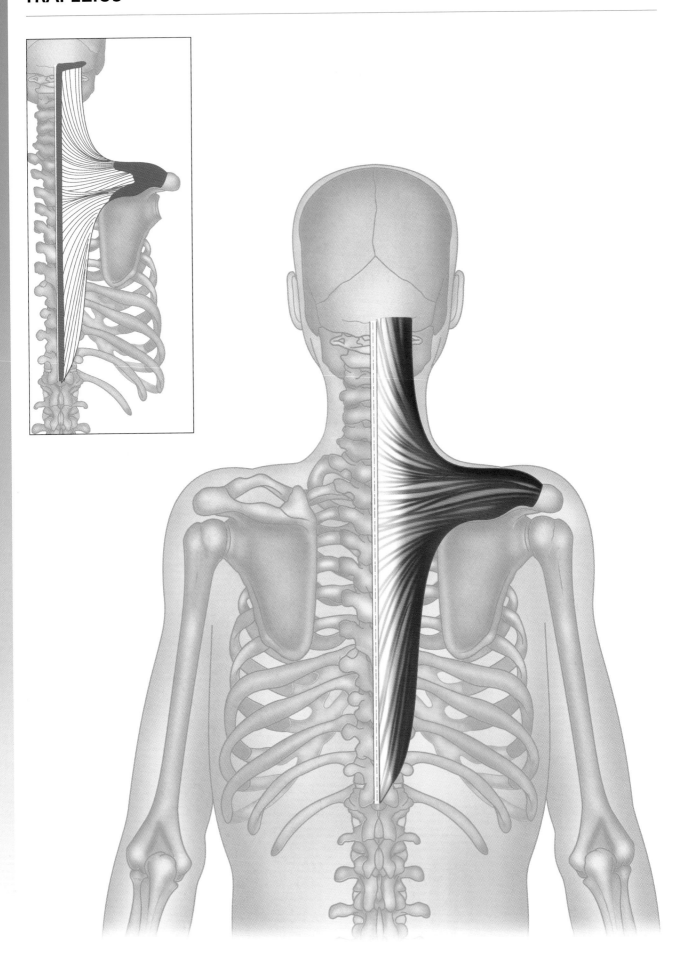

Strengthening exercises

Shoulder press
(upper fibres)

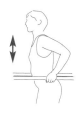

Dips
(middle/lower fibres)

Chin-ups
(middle/lower fibres)

Lateral dumb-bell raises

Greek, *trapezoides*, table shaped.

The left and right trapezius, viewed as a whole, create a trapezium in shape, thus giving this muscle its name.

Origin
Base of skull (occipital bone). Spinous processes of seventh cervical (C7) and all thoracic vertebrae, (T1–T12).

Insertion
Lateral third of clavicle. Acromion process. Spine of scapula.

Action
Upper fibres: Pull the shoulder girdle up (elevation). Helps prevent depression of the shoulder girdle when a weight is carried on the shoulder or in the hand.
Middle fibres: Retract (adduct) scapula.
Lower fibres: Depress scapula, particularly against resistance, as when using the hands to get up from a chair.
Upper and lower fibres together: Rotate scapula, as in elevating the arm above the head.

Nerve
Accessory **X1** nerve. Ventral ramus of cervical nerves, C2, **3**, **4**.

Basic functional movement
Example (upper and lower fibres working together): Painting a ceiling.

Sports that heavily utilise this muscle
Examples: Shot put. Boxing. Seated rowing.

Common problems when muscle is chronically tight / shortened
Upper fibres: Neck pain or stiffness, headaches.

Self stretch

Turn head to right and tuck chin in. Pull left shoulder down. Pull head and left shoulder apart from each other.

LEVATOR SCAPULAE

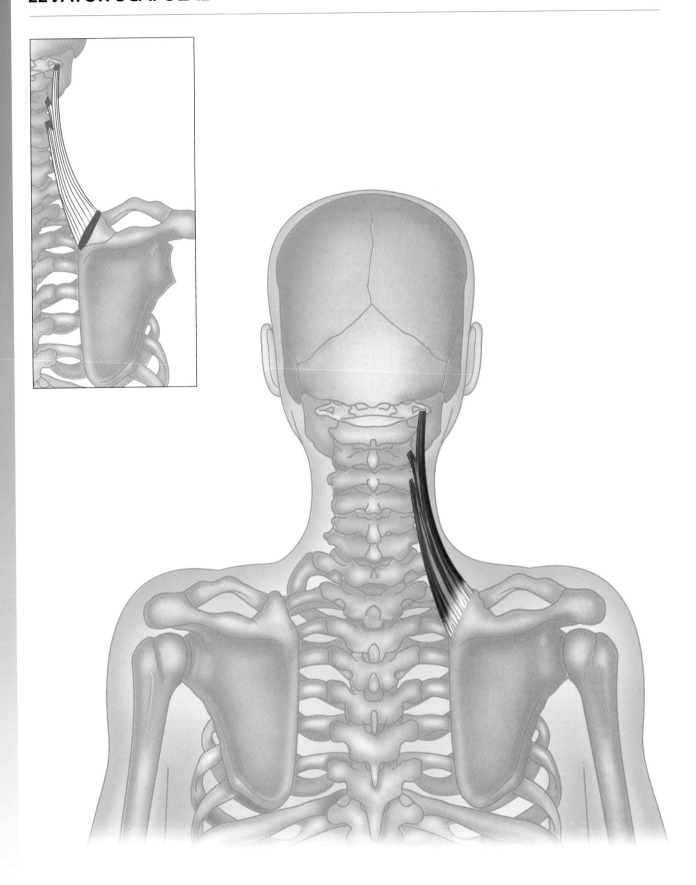

Strengthening exercises

Shrugs with dumb-bells or barbell

Upright (vertical) rowing

Latin, *levare*; to lift; *scapulae*, shoulder, blade(s).

Levator scapulae is deep to sternocleidomastoid and trapezius. It is named after its action of elevating the scapula.

Origin
Transverse processes of the first three or four cervical vertebrae, (C1–C4).

Insertion
Upper medial (vertebral) border of the scapula (i.e. portion above the spine of the scapula).

Action
Elevates scapula. Helps retract scapula. Helps bend neck laterally.

Nerve
Dorsal scapular nerve, C**4**, **5** and cervical nerves, C**3**, **4**.

Basic functional movement
Carrying a heavy bag.

Sports that heavily utilise this muscle
Examples: Shot put. Weight lifting.

Common problems when muscle is chronically tight / shortened
Upper fibres: Neck pain or stiffness, headaches.

Self stretches

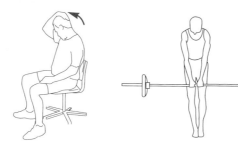

Drop chin to chest and turn chin 45°. Keep spine long.

RHOMBOIDS (MINOR AND MAJOR)

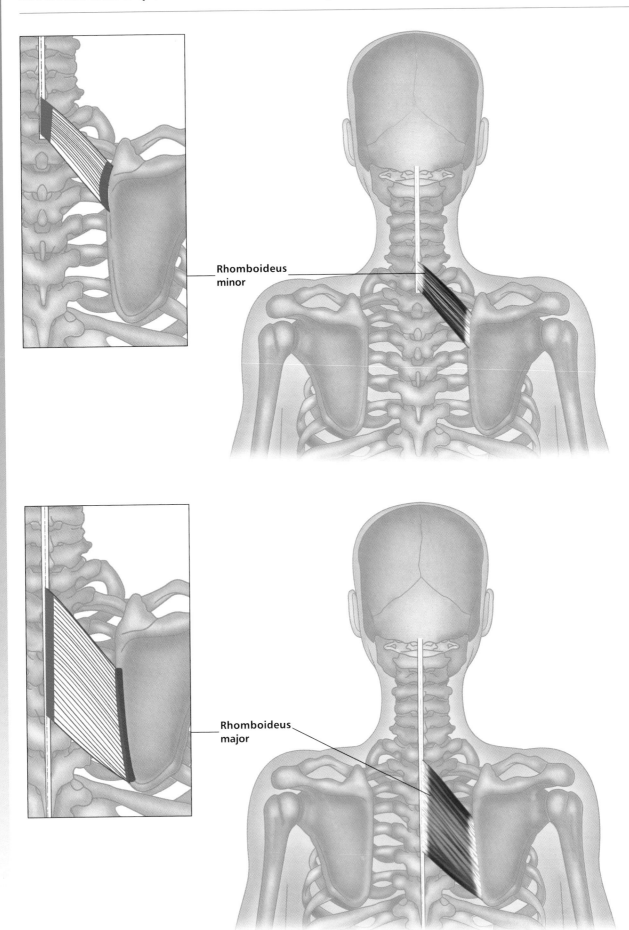

Rhomboideus
minor

Rhomboideus
major

Strengthening exercises

Seated rowing

Upright (vertical) rowing

Lat. pull-downs

Pulley shoulder adduction

Greek, *rhomb*, a parallelogram with oblique angles and only the opposite sides equal; *minor*, small; *major*, large.

Rhomboideus major runs parallel to, and is often continuous with, rhomboideus minor.

Origin
Spinous processes of the seventh cervical and upper five thoracic vertebrae, (C7–T1).

Insertion
Medial (vertebral) border of scapula.

Action
Retracts (adducts) scapula. Stabilizes scapula. Slightly assists in outer range of adduction of arm (i.e. from arm overhead to arm at shoulder level).

Nerve
Dorsal scapular nerve, C**4**, **5**.

Basic functional movement
Pulling something towards you, such as opening a drawer.

Sports that heavily utilise these muscles
Examples: Archery. Seated rowing. Wind surfing. Racket sports.

Common problems when muscles are tight or overstretched
Tight: Soreness or aching between shoulder blades.
Overstretched: Rounded shoulders are both symptomatic of, and exacerbated by, overstretched rhomboids (which tend to get overstretched rather than become too tight).

Self stretch

SERRATUS ANTERIOR

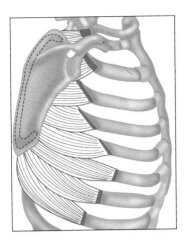

Insertion on anterior of scapula.
Lateral view.

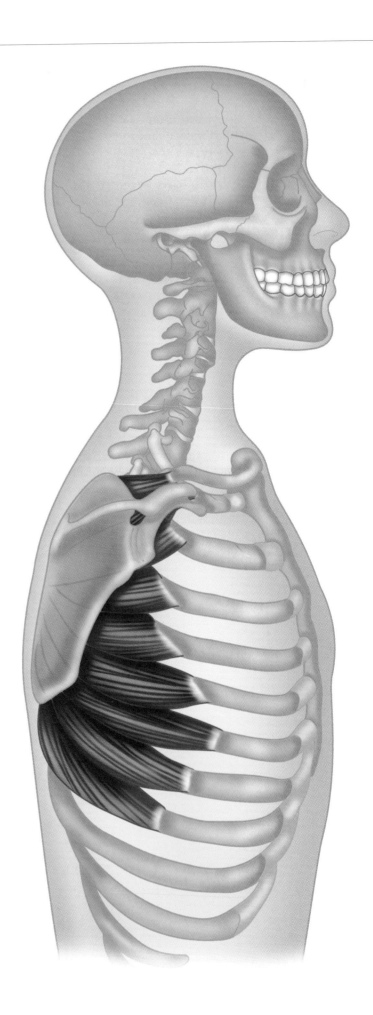

Strengthening exercises

Bench press
(including inclined version)

Shoulder press

Press-ups

Latin, *serratus*, notched; *anterior*, before.

The serratus anterior forms the medial wall of the axilla, along with the upper five ribs. It is a large muscle composed of a series of finger-like slips. The lower slips interdigitate with the origin of the external oblique.

Origin
Outer surfaces and superior borders of upper eight or nine ribs, and the fascia covering their intercostal spaces.

Insertion
Anterior (costal) surface of the medial border of scapula and inferior angle of scapula.

Action
Protracts scapula (pulls it forward on the ribs and holds it closely into the chest wall). Rotates scapula for abduction and flexion of arm.

Nerve
Long thoracic nerve, C**5**, **6**, **7**, 8.

Basic functional movement
Pushing or reaching forwards for something barely within reach.

Sports that heavily utilise this muscle
Examples: Boxing. Shot put.

Common problems when muscle is weak
'Winged scapula' (looking like an angel's wing), especially when holding a weight in front of the body. This is also a feature when the nerve to this muscle is damaged.

Self stretch

PECTORALIS MINOR

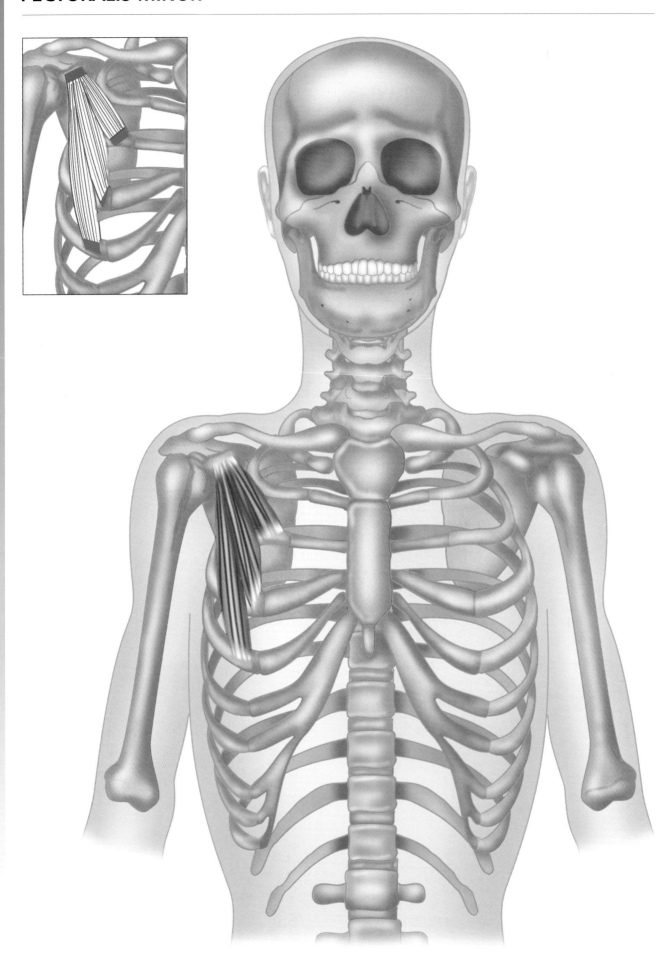

Strengthening exercises

Bench press

Dumb-bell flyes

Pull-overs

Latin, *pectoralis*, chest; *minor*, small.

Pectoralis minor is a flat triangular muscle lying posterior to, and concealed by, pectoralis major. Along with pectoralis major, it forms the anterior wall of the axilla.

Origin
Outer surfaces of third, fourth and fifth ribs and fascia of the corresponding intercostal spaces.

Insertion
Corocoid process of scapula.

Action
Draws scapula forward and downward. Raises ribs during forced inspiration (i.e. it is an accessory muscle of inspiration, if the scapula is stabilized by the rhomboids and trapezius).

Nerve
Medial pectoral nerve with fibres from a communicating branch of the lateral pectoral nerve, C(6), **7**, **8**, T1.

Basic functional movement
Example: Pushing on arms of chair to stand up.

Sports that heavily utilise this muscle
Racket sports, e.g. tennis, badminton. Baseball pitching. Sprinting.

Common problems when muscle is chronically tight / shortened
Restricts expansion of chest.

Self stretches

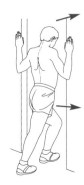

Fix arm against a door frame.
Step forward keeping your back
lengthened, not arched.
Raising or lowering arm will
stretch different parts
of the muscle.

PECTORALIS MAJOR

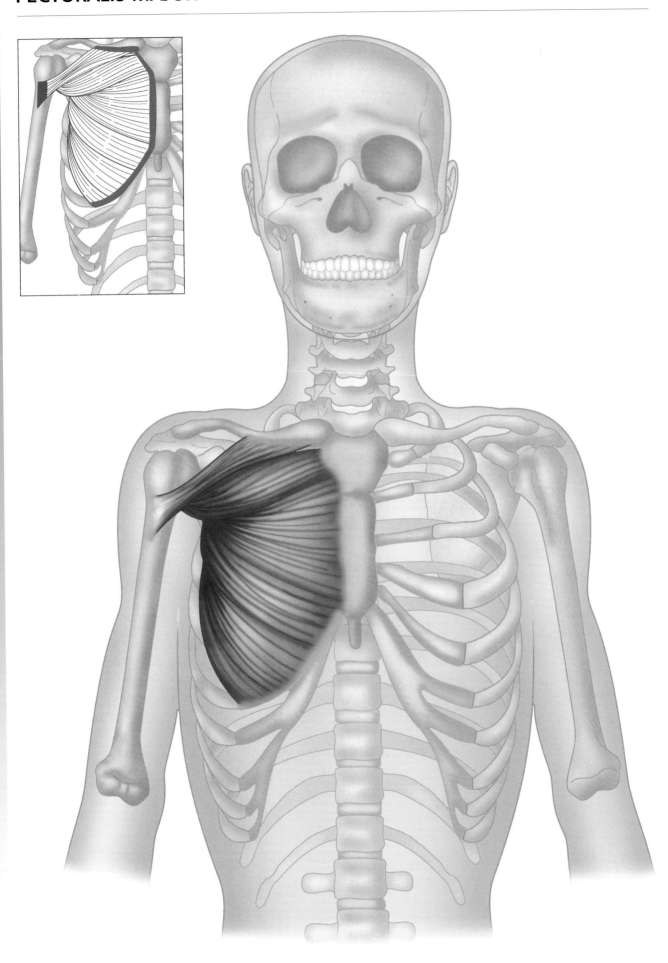

Strengthening exercises

Bench press

Dumb-bell flyes

Vertical flyes ('pec deck' machine / seated butterfly)

Pull-overs

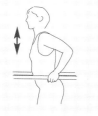

Dips

Latin, *pectoralis*, chest; *major*, large.

Along with pectoralis minor, it forms the anterior wall of the axilla.

Origin
Clavicular head: Medial half or two-thirds of front of clavicle.
Sternocostal portion: Sternum and adjacent upper six costal cartilages.

Insertion
Upper shaft of humerus.

Action
Adducts and medially rotates the humerus.
Clavicular portion: Flexes and medially rotates the shoulder joint, and horizontally adducts the humerus towards the opposite shoulder.
Sternocostal portion: Obliquely adducts the humerus towards the opposite hip.
The pectoralis major is one of the main climbing muscles, pulling the body up to the fixed arm.

Nerve
Nerve to upper fibres: Lateral pectoral nerve, C5, 6, 7.
Nerve to lower fibres: Lateral and medial pectoral nerves, C6, 7, 8, T1.

Basic functional movement
Clavicular portion: Brings arm forwards and across the body, as in applying deodorant to opposite armpit.
Sternal portion: Pulling down from above, such as a rope in bell ringing.

Sports that heavily utilise this muscle
Examples: Racket sports such as tennis. Golf. Baseball pitching. Gymnastics (rings and high bar). Judo. Wrestling.

Movements or injuries that may damage this muscle
Indian wrestling and other strength activities that force medial rotation and adduction can damage the insertion of this muscle.

Common problems when muscle is tight
Rounds the back and restricts expansion of chest, restricting lateral rotation and abduction of the shoulder.

Self stretches

Fix arm against a door frame.
Step forward keeping your back lengthened, not arched.
Raising or lowering arm will stretch different parts of the muscle.

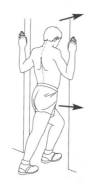

LATISSIMUS DORSI

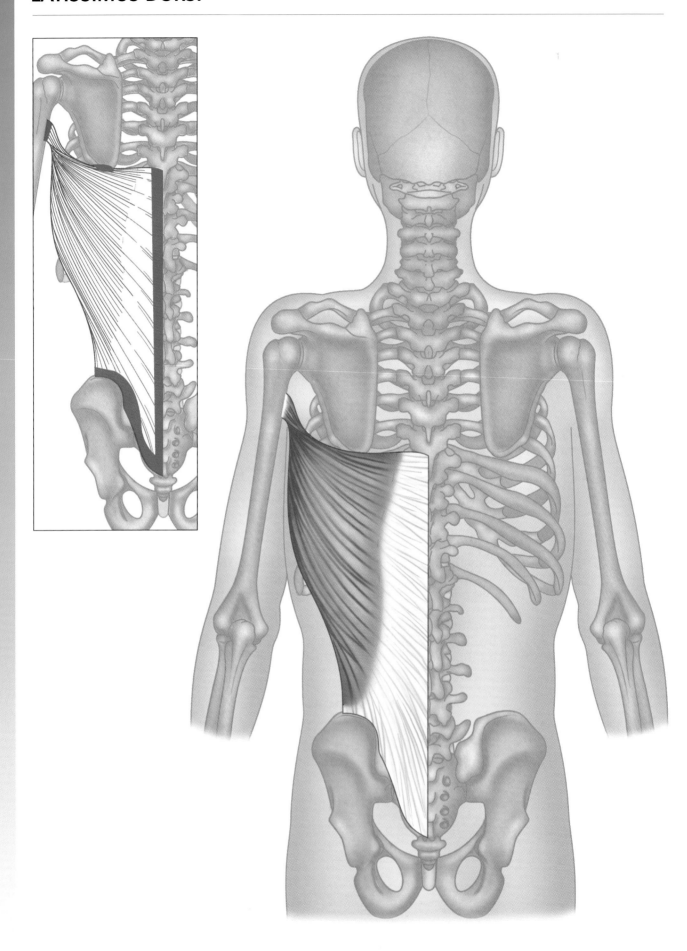

Strengthening exercises

Chin-ups (esp. wide grip)

Lat. pull-downs

Pull-overs

Seated rowing

Pulley shoulder adduction

Latin, *latissimus*, widest; *dorsi*, of the back.

Along with subscapularis and teres major, the latissimus dorsi forms the posterior wall of the axilla.

Origin
A broad sheet of tendon which is attached to the spinous processes of lower six thoracic vertebrae and all the lumbar and sacral vertebrae, (T7–S5). Posterior part of iliac crest. Lower three or four ribs. Inferior angle of the scapula.

Insertion
Twists to insert into the intertubercular sulcus (bicipital groove) of humerus, just below the shoulder joint.

Action
Extends the flexed arm. Adducts and medially rotates the humerus (i.e. draws the arm back and inwards towards the body).

It is one of the chief climbing muscles, since it pulls the shoulders downwards and backwards, and pulls the trunk up to the fixed arms (therefore, also active in crawl swimming stroke). Assists in forced inspiration, by raising the lower ribs.

Nerve
Thoracodorsal nerve, C**6**, **7**, **8**, from the posterior cord of the brachial plexus.

Basic functional movement
Example: Pushing on arms of chair to stand up.

Sports that heavily utilise this muscle
Examples: Climbing. Gymnastics (rings, parallel bars). Swimming. Rowing.

Self stretches

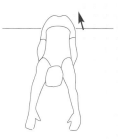

Pull right elbow to left with left hand. Side bending to left will increase stretch.

From kneeling on all fours, sit back onto your ankles, keeping your hands fixed. Relax into it and hold for up to two minutes.

DELTOIDEUS

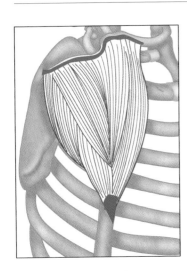

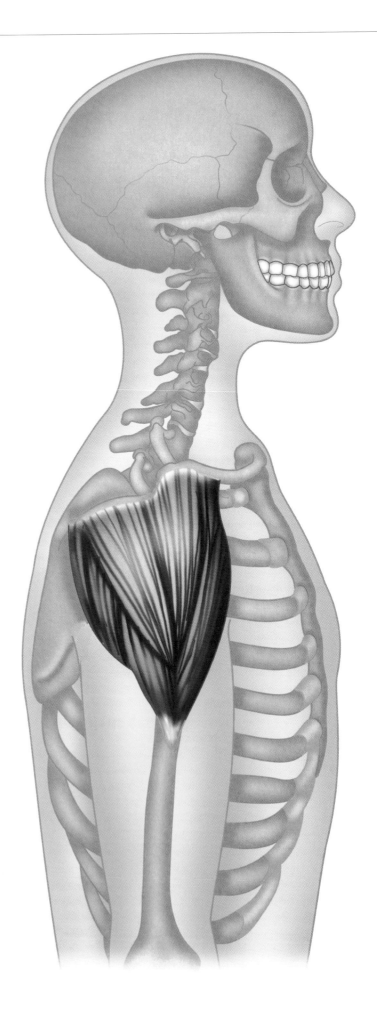

Strengthening exercises

Lateral dumb-bell raises
(middle fibres)

Upright (vertical) rowing
(mainly middle fibres)

Bench press
(anterior fibres)

Shoulder press
(mainly middle fibres)

Vertical flyes ('pec' deck /
seated butterfly)
(anterior fibres)

Greek, *delta*, fourth letter of Greek alphabet (shaped like a triangle).

The deltoid is composed of three parts; anterior, middle and posterior. Only the middle part is multipennate, probably because its mechanical disadvantage of abduction of the shoulder joint requires extra strength.

Origin
Clavicle, acromion process and spine of scapula.

Insertion
Deltoid tuberosity, situated halfway down the lateral surface of the shaft of the humerus.

Action
Anterior fibres: Flex and medially rotate the humerus.
Middle fibres: Abduct the humerus at the shoulder joint (only after the movement has been initiated by supraspinatus).
Posterior fibres: Extend and laterally rotate the humerus.

Nerve
Axillary nerve, C**5**, **6**, from the posterior cord of the brachial plexus.

Basic functional movement
Examples: Reaching for something out to the side, or raising the arm to wave.

Sports that heavily utilise this muscle
Examples: Javelin. Shot put. Racket sports. Wind surfing. Weight lifting.

Self stretches

Keep your arms and torso straight and slowly bend your knees. Vary by placing back of hands on table (mainly for the anterior fibres).

Raise one arm to shoulder height. Flex the arm across to the other shoulder. Hold the raised elbow with the opposite hand and pull the elbow backward.

SUPRASPINATUS

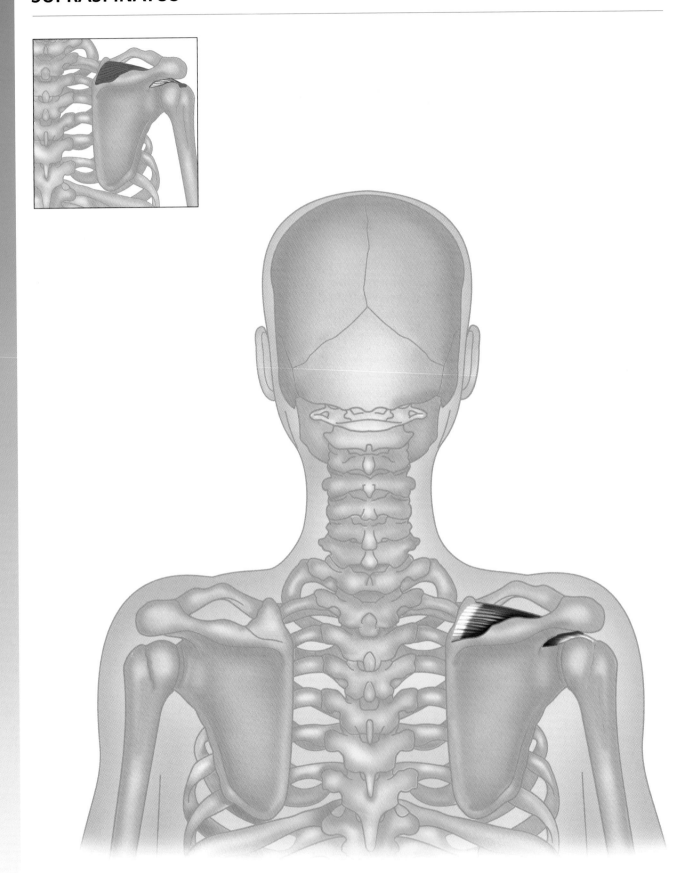

Strengthening exercises

Lateral dumb-bell raises

Seated rowing

Latin, *supra*, above; *spinatus*, spine of the scapula.

A member of the **rotator cuff**, which comprise: supraspinatus, infraspinatus, teres minor, and subscapularis. The rotator cuff helps hold the head of the humerus in contact with the glenoid cavity (socket of shoulder joint) of the scapula during movements of the shoulder, thus helping to prevent dislocation of the joint.

Origin
Supraspinous fossa of scapula (hollow above the spine of the scapula).

Insertion
Greater tubercle at the top of the humerus. Capsule of shoulder joint.

Action
Initiates the process of abduction at the shoulder joint, so that the deltoid can take over at the later stages of abduction.

Nerve
Suprascapular nerve, C4, **5**, 6, from the upper trunk of the brachial plexus.

Basic functional movement
Example: Holding shopping bag away from side of body.

Sports that heavily utilise this muscle
Examples: Baseball. Golf. Racket sports.

Movements or injuries that may damage this muscle
Dislocation of the shoulder joint.

Self stretch

Raise one arm to shoulder height. Flex the arm across to the other shoulder. Hold the raised elbow with the opposite hand and pull the elbow backward.

INFRASPINATUS

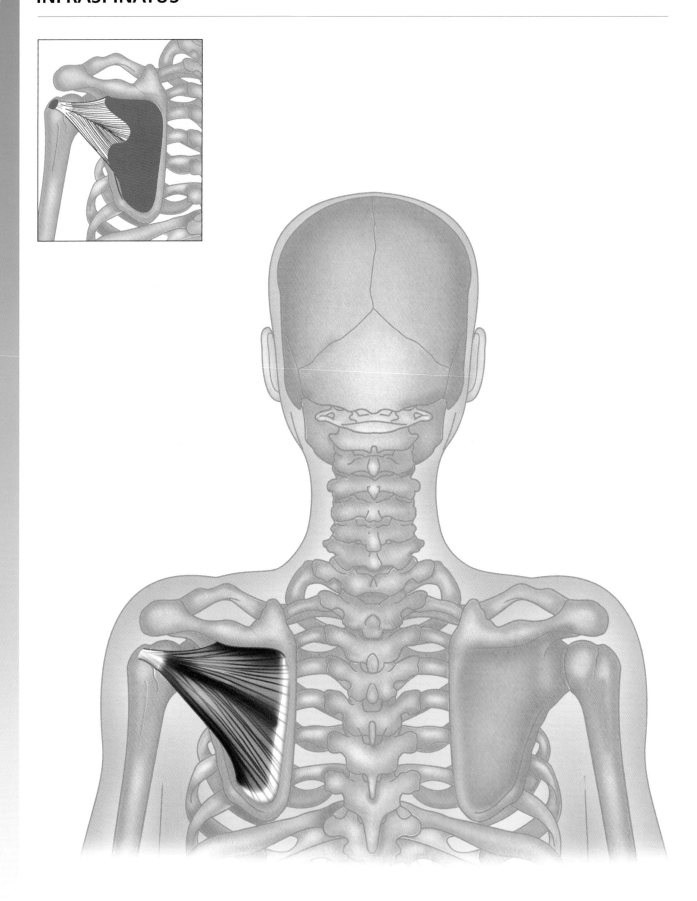

Latin, *infra*, below; *spinatus*, spine of the scapula.

A member of the **rotator cuff**, which comprise: supraspinatus, infraspinatus, teres minor, and subscapularis. The rotator cuff helps hold the head of the humerus in contact with the glenoid cavity (socket of shoulder joint) of the scapula during movements of the shoulder, thus helping to prevent dislocation of the joint.

Origin
Middle two-thirds of dorsal surface of scapula, below spine of scapula.

Insertion
Greater tubercle at the top of humerus. Capsule of shoulder joint.

Action
As a rotator cuff, helps prevent posterior dislocation of the shoulder joint. Laterally rotates humerus.

Nerve
Suprascapular nerve, C(4), **5**, **6**, from the upper trunk of the brachial plexus.

Basic functional movement
Example: Brushing back of hair.

Sports that heavily utilise this muscle
Example: Back hand racket sports.

Movements or injuries that may damage this muscle
Dislocation of the shoulder joint.

Strengthening exercise

Seated rowing
(limited effect)

Self stretches

Hold doorknob and gently step away from door.

Raise one arm to shoulder height. Flex the arm across to the other shoulder. Hold the raised elbow with the opposite hand and pull the elbow backward.

TERES MINOR

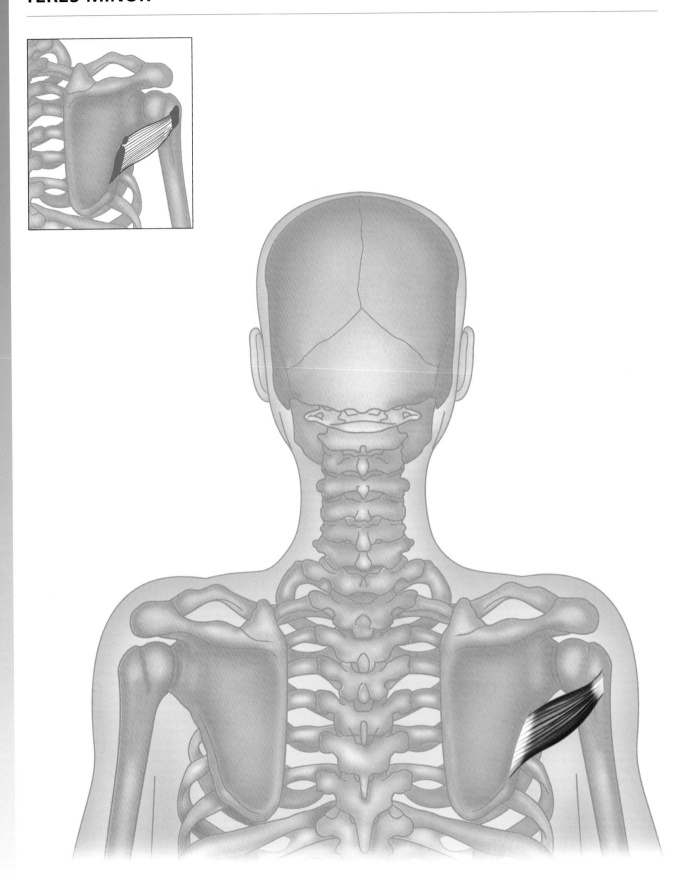

Strengthening exercise

Seated rowing
(limited effect)

Latin, *teres*, rounded, finely shaped; *minor*, small.

A member of the **rotator cuff**, which comprise: supraspinatus, infraspinatus, teres minor, and subscapularis. The rotator cuff helps hold the head of the humerus in contact with the glenoid cavity (socket of shoulder joint) of the scapula during movements of the shoulder, thus helping to prevent dislocation of the joint.

Origin
Upper two-thirds of the lateral edge of the dorsal surface of scapula.

Insertion
Back of greater tubercle of humerus. Capsule of shoulder joint.

Action
As a rotator cuff, helps prevent upward dislocation of the shoulder joint. Laterally rotates humerus. Weakly adducts humerus.

Nerve
Axillary nerve, C**5**, **6**, from the posterior cord of the brachial plexus.

Basic functional movement
Example: Brushing back of hair.

Sports that heavily utilise this muscle
Example: Back hand racket sports.

Movements or injuries that may damage this muscle
Dislocation of the shoulder joint.

Self stretches

Hold doorknob and gently step away from door.

Raise one arm to shoulder height. Flex the arm across to the other shoulder. Hold the raised elbow with the opposite hand and pull the elbow backward.

SUBSCAPULARIS

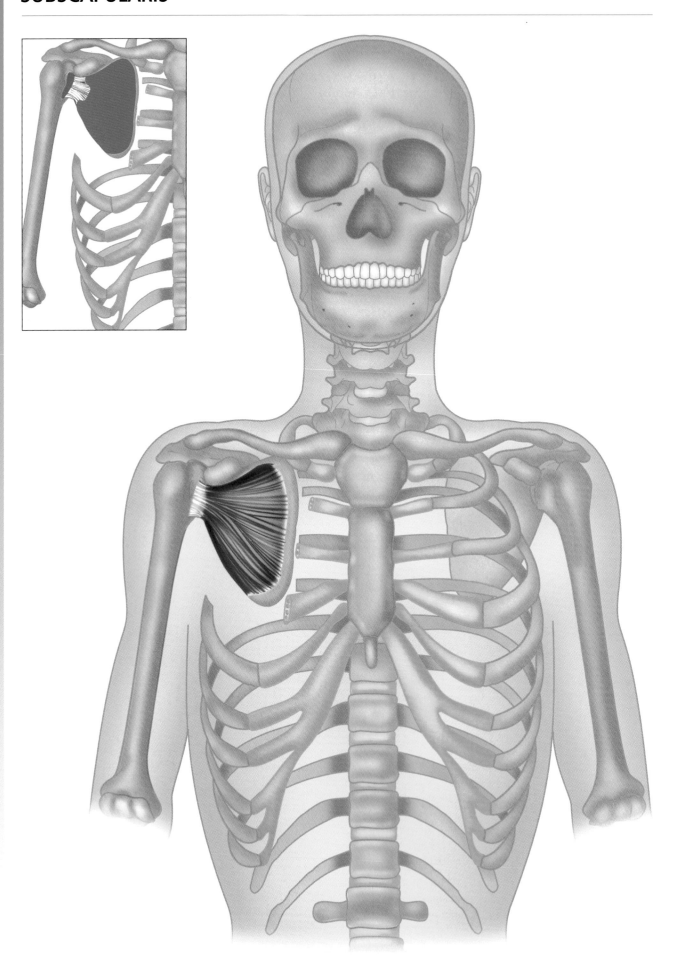

Strengthening exercise

Seated rowing
(limited effect)

Latin, *sub*, under; *scapular*, pertaining to the scapula.

A member of the **rotator cuff**, which comprise: supraspinatus, infraspinatus, teres minor, and subscapularis. The rotator cuff helps hold the head of the humerus in contact with the glenoid cavity (socket of shoulder joint) of the scapula during movements of the shoulder, thus helping to prevent dislocation of the joint. The subscapularis constitutes the greater part of the posterior wall of the axilla.

Origin
Subscapular fossa (anterior surface of scapula).

Insertion
Lesser tubercle at the top of humerus. Capsule of shoulder joint.

Action
As a rotator cuff, stabilizes shoulder joint; mainly prevents the head of the humerus being pulled upwards by the deltoid, biceps brachii and long head of triceps brachii. Medially rotates humerus.

Nerve
Upper and lower subscapular nerves, C**5**, **6**, 7, from the posterior cord of the brachial plexus.

Basic functional movement
Example: Reaching into your back pocket.

Sports that heavily utilise this muscle
Examples: Athletic throwing events. Golf. Racket sports.

Movements or injuries that may damage this muscle
Twisting the arm behind the back, (as in an over zealous restraining hold) or struggling to free oneself from that position, may damage the insertion.

Self stretch

Laterally rotate humerus
with elbow bent 90°,
and anchor hand against
door frame.

TERES MAJOR

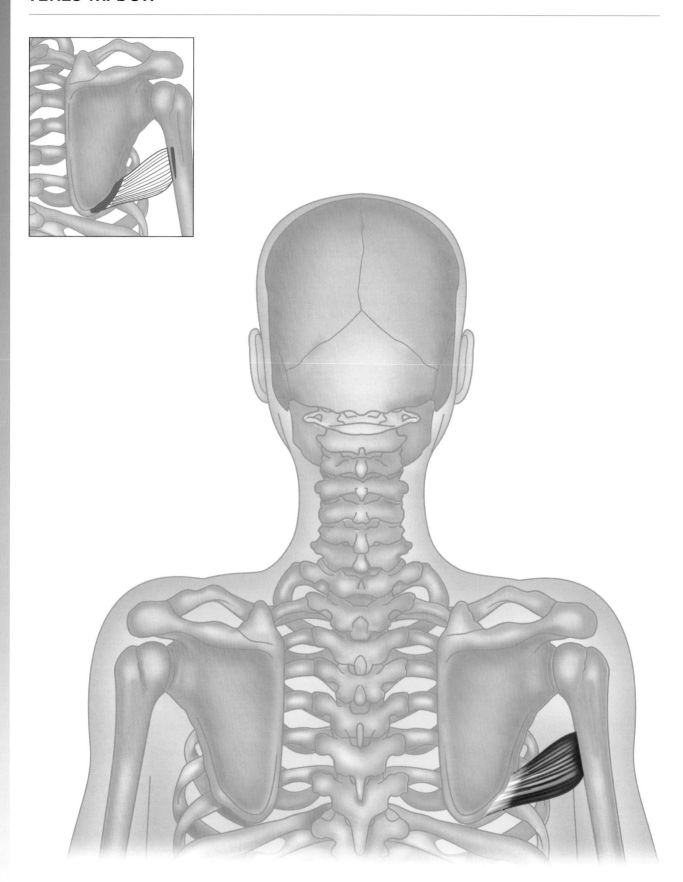

Strengthening exercises

Seated rowing

Pull-overs

Pulley shoulder adduction

Latin, *teres*, rounded, finely shaped; *major*, large.

The teres major, along with the tendon of latissimus dorsi, which passes around it, and the subscapularis, forms the posterior fold of the axilla.

Origin
Lower third of the posterior surface of the lateral border of the scapula.

Insertion
Medial lip of intertubercular sulcus (bicipital groove) of humerus (i.e. back of upper shaft of humerus).

Action
Adducts humerus. Medially rotates humerus. Extends humerus from the flexed position.

Nerve
Lower subscapular nerve, C5, **6**, 7, from the posterior cord of the brachial plexus.

Basic functional movement
Example: Reaching into your back pocket.

Sports that heavily utilise this muscle
Examples: Rowing. Cross-country skiing.

Movements or injuries that may damage this muscle
Sharply jerking the arm forwards, as in throwing a stone to skim it across a lake.

Self stretches

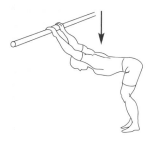

Keep your arms and legs straight, flex the hips and flatten your back.

Raise one arm to shoulder height. Flex the arm across to the other shoulder. Hold the raised elbow with the opposite hand and pull the elbow backward.

BICEPS BRACHII

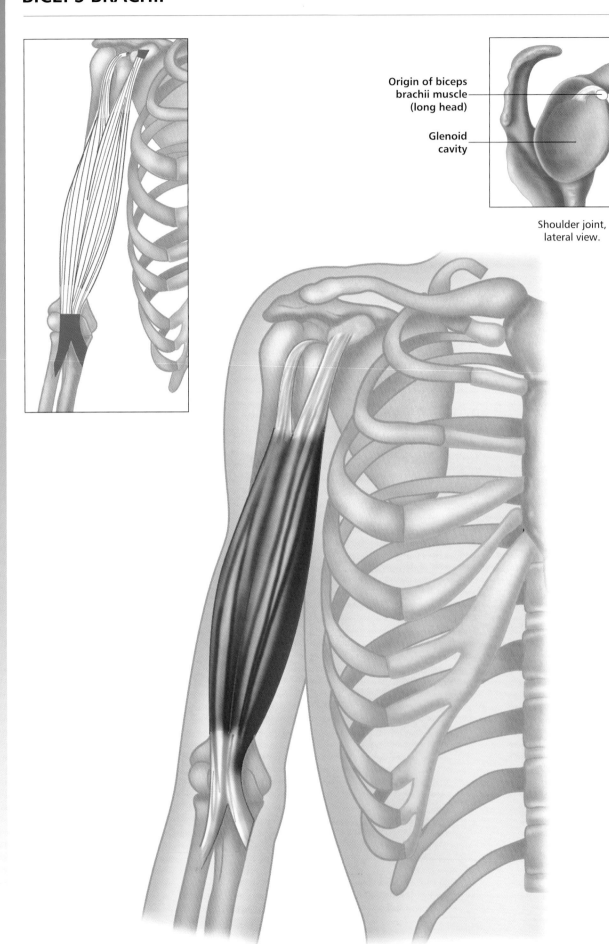

Origin of biceps brachii muscle (long head)

Glenoid cavity

Shoulder joint, lateral view.

Strengthening exercises

Biceps curl

Chin-ups

Lat. pull downs

Latin, *biceps*, two-headed muscle; *brachii*, of the arm.

Biceps brachii operates over three joints. It has two tendinous heads at its origin and two tendinous insertions. Occasionally it has a third head, originating at the insertion of coracobrachialis. The short head forms part of the lateral wall of the axilla, along with coracobrachialis and the humerus.

Origin
Short head: Tip of corocoid process of scapula.
Long head: Supraglenoid tubercle of scapula (area just above socket of shoulder joint).

Insertion
Radial tuberosity (on medial aspect of upper part of shaft of radius). Deep fascia (connective tissue) on medial aspect of forearm.

Action
Flexes elbow joint. Supinates forearm. (It has been described as the muscle that puts in the corkscrew and pulls out the cork). Weakly flexes arm at the shoulder joint.

Nerve
Musculocutaneous nerve, C**5**, **6**.

Basic functional movement
Examples: Picking up an object. Bringing food to mouth.

Sports that heavily utilise this muscle
Examples: Boxing. Climbing. Canoeing. Rowing.

Movements or injuries that may damage this muscle
Lifting heavy objects too suddenly.

Common problems when muscle is chronically tight / shortened
Flexion deformity of elbow (elbow cannot be fully straightened).

Self stretches

CORACOBRACHIALIS

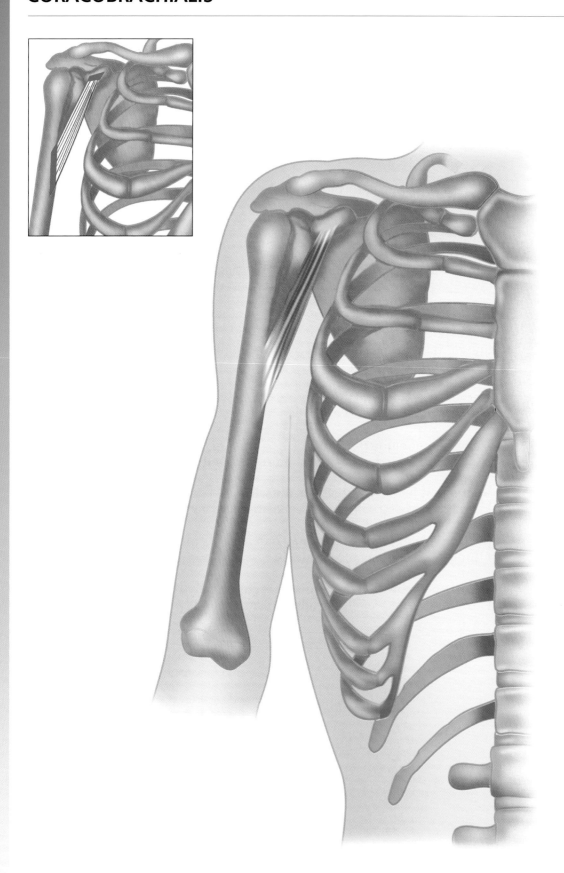

Strengthening exercise

Pulley shoulder adduction

Greek, *coracoid*, raven's beak; **Latin**, *brachial*; relating to the arm.

Along with the short head of biceps brachii and the humerus, the coracobrachialis forms the lateral wall of the axilla. Sometimes known as the 'armpit' muscle.

Origin
Tip of the corocoid process of scapula.

Insertion
Medial aspect of humerus at mid-shaft.

Action
Weakly adducts shoulder joint. Possibly assists in flexion of the shoulder joint (but this has not been proven). Helps stabilize humerus.

Nerve
Musculocutaneous nerve, C6, 7.

Basic functional movement
Example: Mopping the floor.

Sports that heavily utilise this muscle
Examples: Golf. Cricket batting.

Movements or injuries that may damage this muscle
Suddenly hitting the ground when swinging the bat hard in cricket.

BRACHIALIS

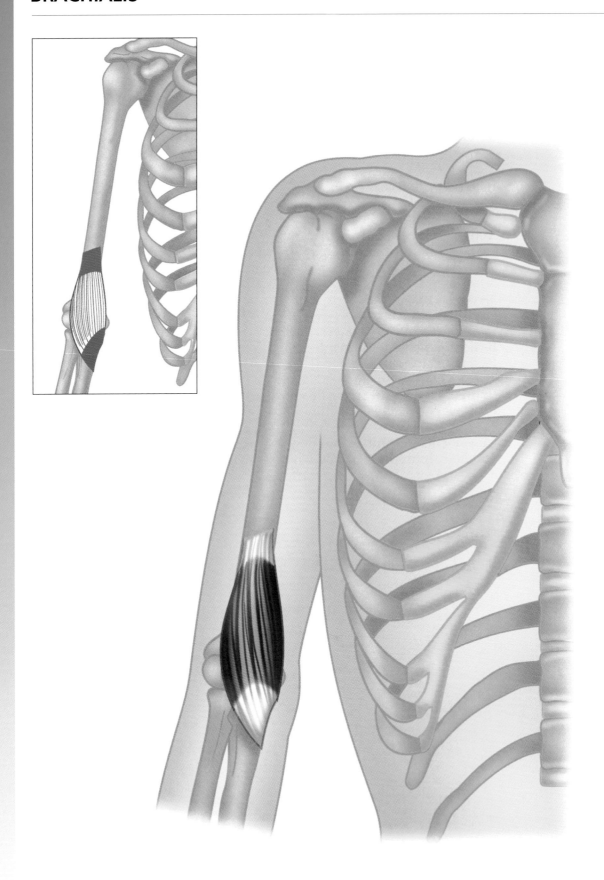

Latin, *brachial*, relating to the arm.

Brachialis lies posterior to biceps brachii and is the main flexor of the elbow joint. Some fibres may be partly fused with the brachioradialis.

Origin
Anterior of lower shaft of humerus.

Insertion
Coronoid process and tuberosity of ulna (i.e. area on front of upper part of shaft of ulna).

Action
Flexes elbow joint.

Nerve
Musculocutaneous nerve, C**5**, **6**.

Basic functional movement
Example: Bringing food to the mouth.

Sports that heavily utilise this muscle
Examples: Baseball. Boxing. Gymnastics.

Common problems when muscle is chronically tight / shortened
Flexion deformity of elbow (elbow cannot be fully straightened).

Strengthening exercises

Biceps curl

Chin-ups

Self stretch

TRICEPS BRACHII

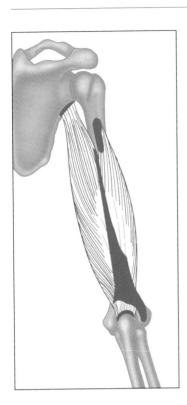

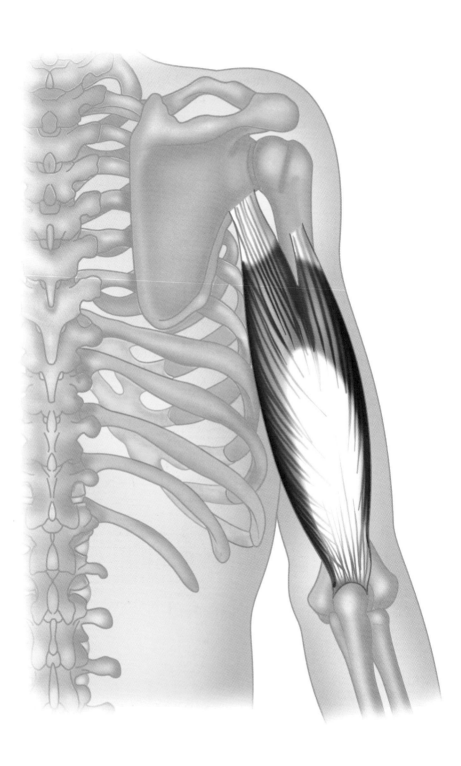

Strengthening exercises

Bench press

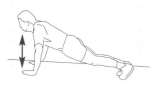

Press-ups

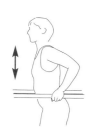

Dips

Triceps kick-back

Latin, *triceps*, three-headed muscle; *brachii*, of the arm.

The triceps originates from three heads and is the only muscle on the back of the arm.

Origin
Long head: Infraglenoid tubercle of the scapula (area just below socket of shoulder joint).
Lateral head: Upper half of posterior surface of shaft of humerus.
Medial head: Lower half of posterior surface of shaft of humerus.

Insertion
Olecranon process of the ulna (i.e. upper posterior area of ulna, near the point of the elbow).

Action
Extends (straightens) elbow joint. Long head can adduct the humerus and extend it from the flexed position. Stabilizes shoulder joint.

Nerve
Radial nerve, C6, **7**, **8**, T1.

Basic functional movement
Examples: Throwing objects. Pushing a door shut.

Sports that heavily utilise this muscle
Examples: Basketball or netball (shooting). Shot put. Baseball (pitcher). Volleyball.

Movements or injuries that may damage this muscle
Throwing with excessive force.

Problems when muscle is chronically tight / shortened
Extension deformity of elbow (elbow cannot be fully flexed); although not very common.

Self stretches

Keep your head up and elbow as far back as is comfortable, without hollowing your lower back.

Pull your hands towards each other. Most effective when the raised elbow is against a wall.

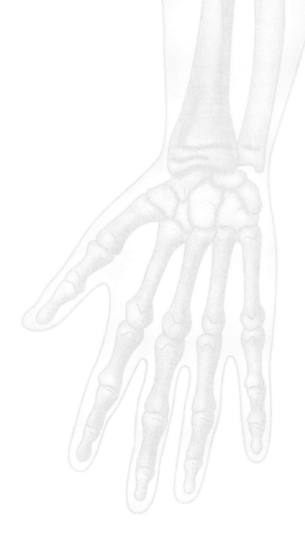

5
Muscles of the Forearm and Hand

PRONATOR TERES

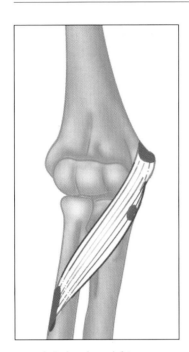

Anterior view, right arm.

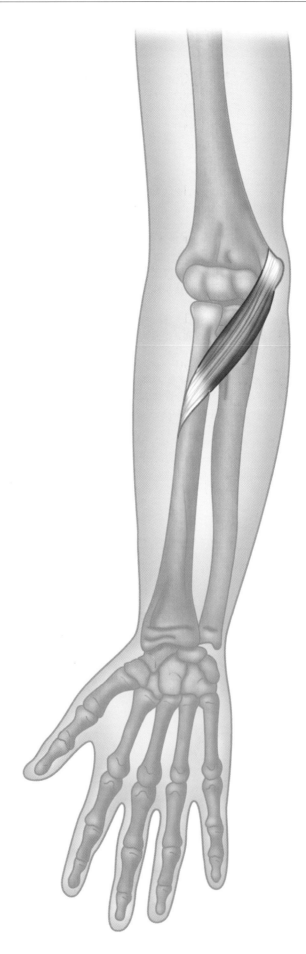

Strengthening exercise

Pronation with strength bar

Latin, *pronate*, bent forward; *teres*, rounded, finely shaped.

Origin
Humeral head: Common flexor origin on the anterior aspect of the medial epicondyle of humerus, and area immediately above (i.e. lower medial end of humerus).
Ulnar head: Coronoid process of the ulna (i.e. area on front of upper part of shaft of ulna).

Insertion
Middle of lateral surface of radius.

Action
Pronates forearm. Assists flexion of elbow joint.

Nerve
Median nerve, C**6**, **7**.

Basic functional movement
Examples: Pouring liquid from a container. Turning a doorknob.

Sports that heavily utilise this muscle
Examples: Batting in cricket. Hockey dribbling. Volleyball smash.

Self stretch

Weight of stick increases supination via gravity.

WRIST FLEXORS

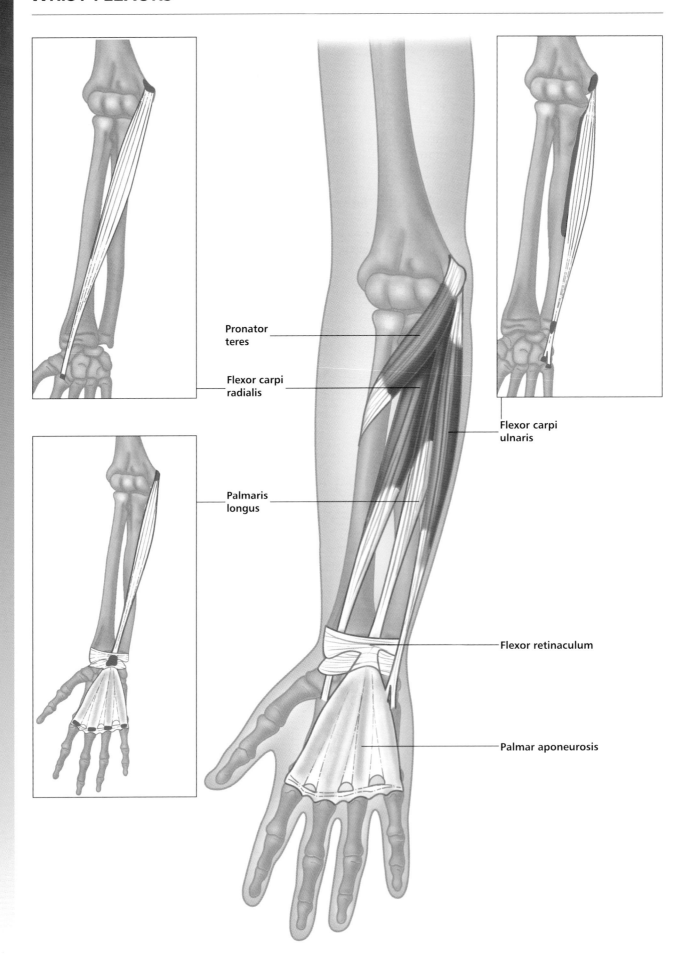

Pronator teres

Flexor carpi radialis

Flexor carpi ulnaris

Palmaris longus

Flexor retinaculum

Palmar aponeurosis

Strengthening exercises

Biceps curl

Wrist rolling (palm up)

Wrist curl

Latin, *flex*, to bend.

Includes: flexor carpi radialis, palmaris longus, flexor carpi ulnaris.

Origin
Common flexor origin on the anterior aspect of the medial epicondyle of humerus (i.e. lower medial end of humerus).

Insertion
Carpals, metacarpals and phalanges.

Action
Flex the wrist (flexor carpi radialis also abducts the wrist; flexor carpi ulnaris also adducts the wrist).

Nerve
Flexor carpi radialis: Median nerve, C**6**, **7**, 8.
Palmaris longus: Median nerve, C(6), **7**, **8**, T1.
Flexor carpi ulnaris: Ulnar nerve, C7, **8**, T1.

Basic functional movement
Examples: Pulling rope in towards you. Wielding an axe or hammer.

Sports that heavily utilise these muscles
Examples: Sailing. Water skiing. Golf. Baseball. Cricket. Volleyball.

Movements or injuries that may damage these muscles
Overextending the wrist resulting from breaking a fall with the hand.

Common problems when muscles are chronically tight / shortened / overused
Golfer's elbow (overuse tendonitis of common flexor origin), carpal tunnel syndrome.

Self stretches

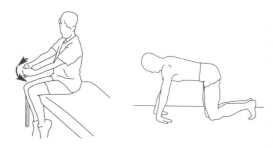

Use one hand to gently lever the other wrist into extension.

FINGER FLEXORS

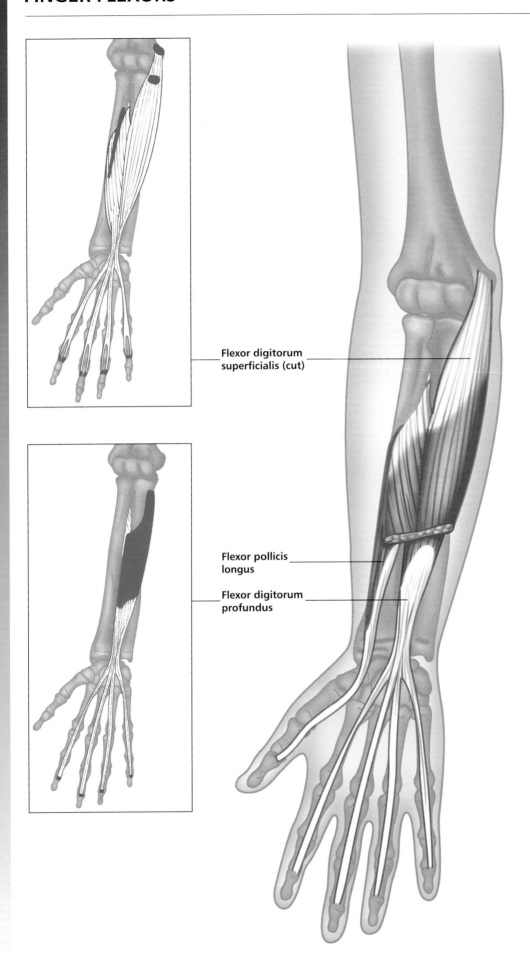

Flexor digitorum
superficialis (cut)

Flexor pollicis
longus

Flexor digitorum
profundus

Strengthening exercises

Biceps curl

Chin-ups

Exer. ring finger flexion

Latin, *flex*, to bend.

Includes flexor digitorum superficialis and flexor digitorum profundus.

Origin
Superficialis: Common flexor tendon on medial epicondyle of humerus. Coronoid process of ulna. Anterior border of radius.
Profundus: Medial and anterior surfaces of the ulna.

Insertion
Superficialis: Sides of the middle phalanges of the four fingers.
Profundus: Base of distal phalanges.

Action
Superficialis: Flexes the middle phalanges of each finger. Can help flex the wrist.
Profundus: Flexes distal phalanges (the only muscle able to do so).

Nerve
Superficialis: Median nerve, C**7**, **8**, T**1**.
Profundus: Medial half of muscle, ulnar nerve, C**7**, **8**, T**1**.
Lateral half of muscle, median nerve, C**7**, **8**, T**1**.
Sometimes the ulnar nerve supplies the whole muscle.

Basic functional movement
Examples: 'Hook grip', as in carrying a briefcase. 'Power grip', as in turning a tap. Typing. Playing the piano and some stringed instruments.

Sports that heavily utilise these muscles
Examples: Archery. Maintaining grip in racket and batting sports. Judo. Rowing. Rock-face climbing.

Movements or injuries that may damage these muscles
Overextending the wrist resulting from breaking a fall with the hand.

Common problems when muscles are chronically tight / shortened / overused
Golfer's elbow (overuse tendonitis of common flexor origin). Carpal tunnel syndrome.

Self stretch

Gently pull each finger
in turn into extension.

BRACHIORADIALIS

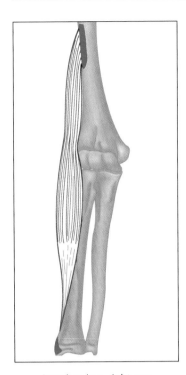

Anterior view, right arm.

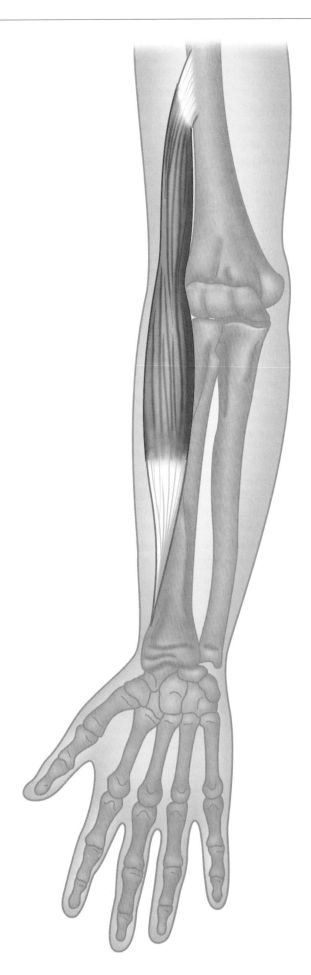

Strengthening exercises

Biceps curl

Chin-ups

Upright rowing

Latin, *brachial*, relating to the arm; *radius*, staff, spoke of wheel.

The brachioradialis forms the lateral border of the cubital fossa. The muscle belly is prominent when working against resistance. Part of the superficial group.

Origin
Upper two-thirds of the anterior aspect of lateral supracondylar ridge of humerus (i.e. lateral part of shaft of humerus, 5–7.5cms (2–3") above elbow joint).

Insertion
Lower lateral end of radius, just above the styloid process.

Action
Flexes elbow joint. Assists in pronating and supinating forearm when these movements are resisted.

Nerve
Radial nerve, C**5**, **6**.

Basic functional movement
Example: Turning a cork screw.

Sports that heavily utilise this muscle
Examples: Baseball. Cricket. Golf. Racket sports. Rowing.

Self stretch

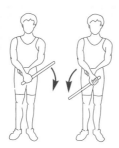

Pronate and
supinate forearm.

SUPINATOR

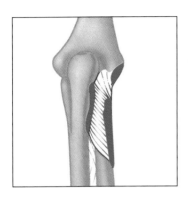

Posterior view, right arm.

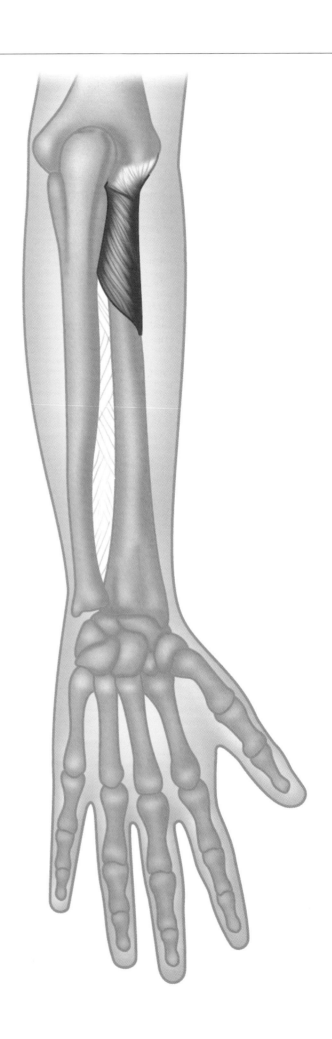

Strengthening exercise

Dumb-bell biceps curl

Latin, *supinus*, lying on the back.

Part of the deep group. Supinator is almost entirely concealed by the superficial muscles.

Origin
Lower lateral end of humerus (lateral epicondyle) and upper lateral end of ulna, and associated ligaments.

Insertion
Dorsal and lateral surfaces of upper third of radius.

Action
Supinates forearm.

Nerve
Deep radial nerve, C5, **6**, (7).

Basic functional movement
Examples: Turning a door handle, or screwdriver.

Sports that heavily utilise this muscle
Example: Backhand in racket sports.

Self stretch

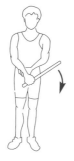

Weight of stick increases pronation via gravity.

WRIST EXTENSORS

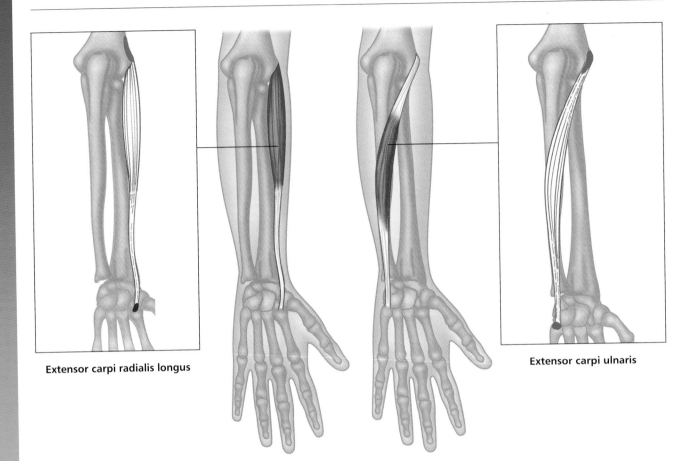

Extensor carpi radialis longus

Extensor carpi ulnaris

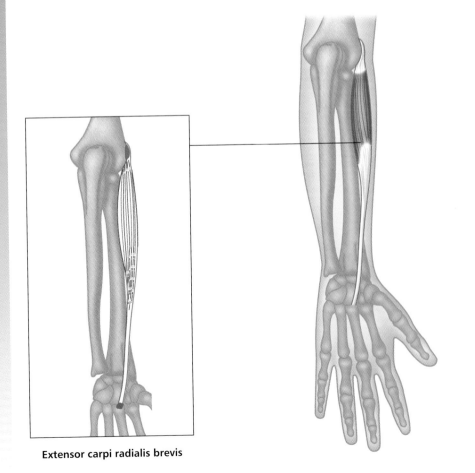

Extensor carpi radialis brevis

Strengthening exercises

Wrist roller (palm down)

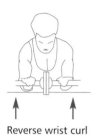

Reverse wrist curl

Most dumb-bell exercises

Latin, *extensor*, to extend.

Includes extensor carpi radialis longus and brevis, and extensor carpi ulnaris.

Origin
Common extensor tendon from lateral epicondyle of humerus (i.e. lower lateral end of humerus).

Insertion
Dorsal surface of metacarpal bones.

Action
Extends the wrist (extensor carpi radialis longus and brevis also abduct the wrist; extensor carpi ulnaris also adducts the wrist).

Nerve
Radialis longus and brevis: Radial nerve, C5, **6**, **7**, 8.
Extensor carpi ulnaris: Deep radial (posterior interosseous) nerve, C6, **7**, **8**.

Basic functional movement
Examples: Kneading dough. Typing. Cleaning windows.

Sports that heavily utilise these muscles
Examples: Back hand badminton. Golf. Motorcycle sports (throttle control).

Movements or injuries that may damage these muscles
Overflexing the wrist resulting from falling onto the hand.

Common problems when muscles are chronically tight / shortened / overused
Tennis elbow (overuse tendonitis of common origin on lateral epicondyle of humerus).

Self stretches

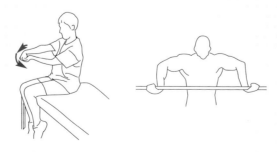

Use lower hand to gently
lever the other wrist
into flexion.

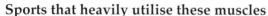

FINGER EXTENSORS (EXTENSOR DIGITORUM)

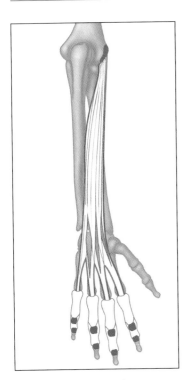

Posterior view, right arm.

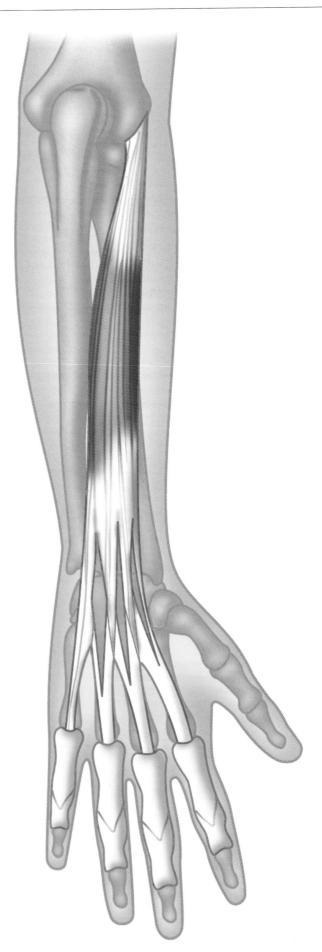

Strengthening exercise

Exer. ring finger extension

Latin, *extensor*, to extend; *digit*, finger.

Origin
Common extensor tendon from lateral epicondyle of humerus (i.e. lower lateral end of humerus).

Insertion
Dorsal surfaces of all the phalanges of the four fingers.

Action
Extends the fingers. Assists abduction (divergence) of fingers away from the middle finger.

Nerve
Deep radial (posterior interosseous) nerve, C**6**, **7**, **8**.

Basic functional movement
Example: Letting go of objects held in the hand.

Movements or injuries that may damage this muscle
Overflexing the wrist resulting from falling onto the hand.

Common problems when muscle is chronically tight / shortened / overused
Tennis elbow (overuse tendonitis of common origin on lateral epicondyle of humerus).

Self stretch

Use one hand to gently lever the other wrist, and therefore fingers, into extension.

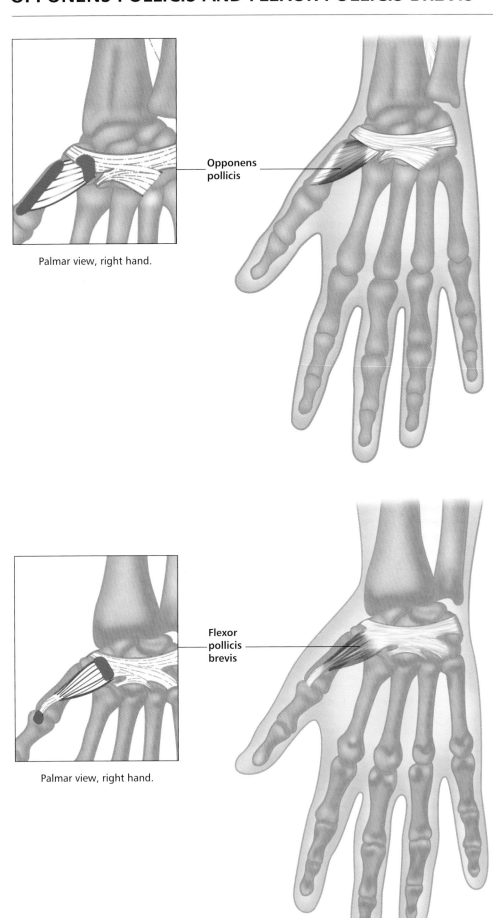

Opponens
pollicis

Palmar view, right hand.

Flexor
pollicis
brevis

Palmar view, right hand.

Strengthening exercises

Exer. ring
'pinching' exercise

Latin, *opponens*, opposing; *pollicis*, of the thumb; *flexor*, to flex; *brevis*, short.

Both part of the thenar eminence, with opponens pollicis usually partly fused with flexor pollicis brevis, and deep to abductor pollicis brevis.

Origin
Opponens pollicis: Flexor retinaculum. Tubercle of trapezium.
Flexor pollicis brevis: *Superficial head*: Flexor retinaculum. Trapezium.
Deep head: Trapezoid. Capitate.

Insertion
Opponens pollicis: Entire length of radial border of first metacarpal.
Flexor pollicis brevis: Radial side of base of proximal phalanx of thumb.

Action
Opponens pollicis: Opposes (i.e. abducts, then slightly medially rotates, followed by flexion and adduction) the thumb so that the pad of the thumb can be drawn into contact with the pads of the fingers.
Flexor pollicis brevis: Flexes the metacarpophalangeal and carpometacarpal joints of the thumb. Assists in opposition of the thumb towards the little finger.

Nerve
Median nerve, (C6, 7, 8, T1).

Basic functional movement
Opponens pollicis: e.g. Picking up small object between thumb and fingers.
Flexor pollicis brevis: e.g. Holding a thread between thumb and fingertips.

Sports that heavily utilise these muscles
Examples: Rock-face climbing. Motorcycle sports (clutch and throttle movement).

Movements or injuries that may damage these muscles
Overextending the thumb resulting from falling on the hand (rare).

Self stretches

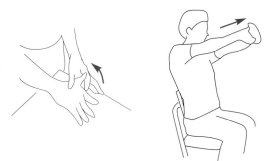

Gently pull thumb
into extension.

LUMBRICALES

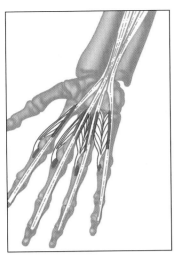

Palmar view, right hand.

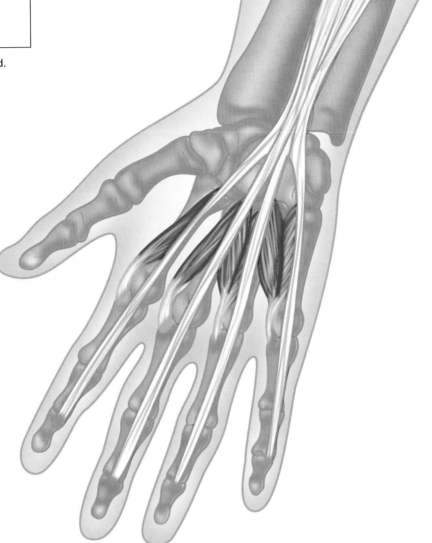

Strengthening exercise

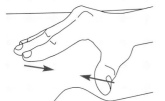

Latin, *lumbrical*, earthworm.

Four small cylindrical muscles, one for each finger, named after the earthworm, because of their shape.

Origin
Tendons of flexor digitorum profundus in the palm.

Insertion
Lateral (radial) side of corresponding tendon of extensor digitorum, on the dorsum of the respective digits.

Action
Extend the interphalangeal joints and simultaneously flex the metacarpophalangeal joints of the fingers.

Nerve
This varies. The usual configuration is:
Lateral lumbricales (first and second): Median nerve, C(6), 7, **8**, T**1**.
Medial lumbricales (third and fourth): Ulnar nerve, C(7), **8**, T**1**.
However, the number of lumbricales supplied by the ulnar nerve may be increased to four or decreased to one.

Basic functional movement
Example: Cupping your hand.

Sports that heavily utilise these muscles
Examples: Volleyball. Handball.

Common problems when muscles are chronically tight / shortened
Clawed hand. Inability to maintain flexion of the interphalangeal joints, as in rock climbing.

Self stretch

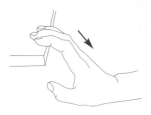

PALMAR INTEROSSEI AND DORSAL INTEROSSEI

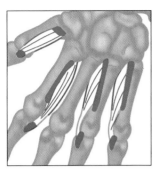

Palmar view, right hand.

Palmar interossei

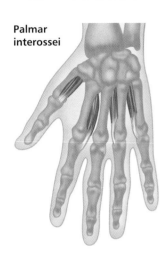

Palmar view, right hand.

Dorsal interossei

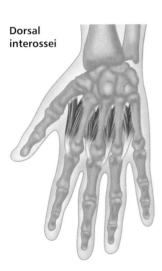

Latin, *palmaris*, palma, palm; *dorsal*, back; *interosseus*, between bones.

The four palmar interossei are located in the spaces between the metacarpals. Each muscle arises from the metacarpal of the digit upon which it acts. The four dorsal interossei are about twice the size of the palmar interossei.
NOTE: The palmar interosseous of the thumb is usually absent.

Origin
Palmar interossei: First: medial (ulnar) side of base of first metacarpal.
Second: medial (ulnar) side of shaft of second metacarpal.
Third: lateral (radial) side of shaft of fourth metacarpal.
Fourth: lateral (radial) side of shaft of fifth metacarpal.
Dorsal interossei: By two heads, each from adjacent sides of metacarpals. Therefore, each dorsal interossei occupies an interspace between adjacent metacarpals.

Insertion
Palmar interossei: Primarily into the extensor expansion of the respective digit, with possible attachment to base of proximal phalanx as follows:
First: Medial (ulnar) side of proximal phalanx of thumb.
Second: Medial (ulnar) side of proximal phalanx of index finger.
Third: Lateral (radial) side of proximal phalanx of ring finger.
Fourth: Lateral (radial) side of proximal phalanx of little finger.
Dorsal interossei: Into the extensor expansion and to base of proximal phalanx as follows:
First: Lateral (radial) side of index finger, mainly to base of proximal phalanx.
Second: Lateral (radial) side of middle finger.
Third: Medial (ulnar) side of middle finger, mainly into extensor expansion.
Fourth: Medial (ulnar) side of ring finger.

Action
Adduct (converge) fingers and thumb towards the middle (third) finger (palmar interossei). Abduct fingers away from middle finger (dorsal interossei). Assist in flexion of fingers at metacarpophalangeal joints.

Nerve
Ulnar nerve, C8, T1.

Basic functional movement
Palmar interossei: e.g. Cupping hand as if to retain water in the palm (i.e. drinking from the hand).
Dorsal interossei: e.g. Spreading fingers, as if to indicate numbers from two to four.

Sport that heavily utilises these muscles
Example: Rock-face climbing.

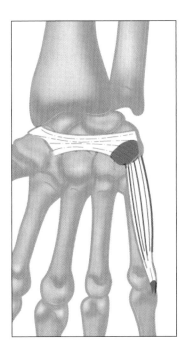

Palmar view, right hand.

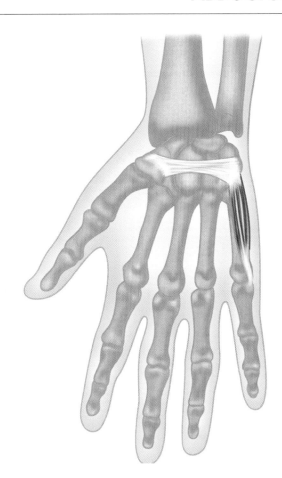

Latin, *abductor*, away from; *digit*, finger; *minimi*, smallest.

This is the most superficial muscle of the hypothenar eminence. The others are flexor digiti minimi brevis and opponens digiti minimi.

Origin
Pisiform bone. Tendon of flexor carpi ulnaris.

Insertion
Ulnar (medial) side of base of proximal phalanx of little finger.

Action
Abducts the little finger. A surprisingly powerful muscle, which particularly comes into play when fingers are spread to grasp a large object.

Nerve
Ulnar nerve, C(7), **8**, T**1**.

Basic functional movement
Example: Holding a large ball.

Sports that heavily utilise this muscle
Examples: Rock-face climbing. Basketball. Netball.

OPPONENS DIGITI MINIMI AND FLEXOR DIGITI MINIMI BREVIS

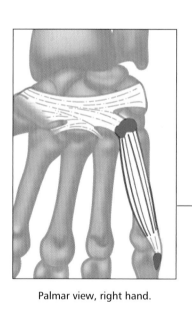

Palmar view, right hand.

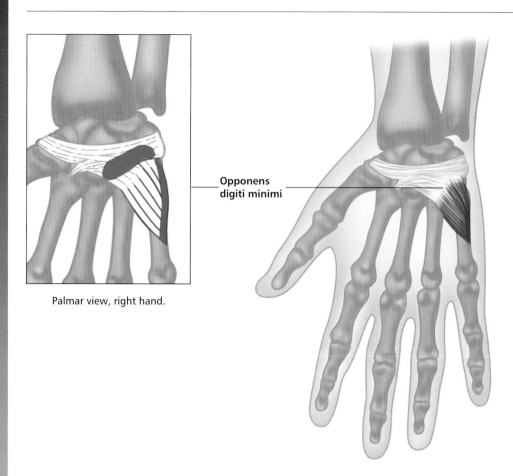

Opponens digiti minimi

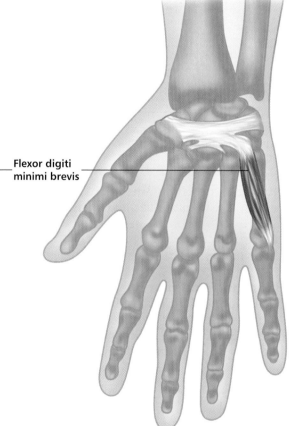

Palmar view, right hand.

Flexor digiti minimi brevis

Strengthening exercises

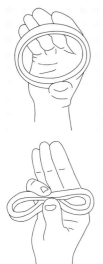

Exer. ring
'pinching' exercise

Latin, *opponens*, opposing; *digit*, finger; *minimi*, smallest; *flexor*, to flex; *brevis*, short.

Both part of the hypothenar eminence, with opponens digiti minimi lying deep to abductor digiti minimi. Flexor digiti minimi brevis may be absent or fused with a neighbouring muscle.

Origin
Hook of hamate. Anterior surface of flexor retinaculum.

Insertion
Opponens digiti minimi: Entire length of medial (ulnar) border of fifth metacarpal.
Flexor digiti minimi brevis: Ulnar (medial) side of base of proximal phalanx of little finger.

Action
Opponens digiti minimi: Pulls metacarpal of the little finger forward and rotates it laterally, so deepening the hollow of the hand, and enabling the pad of the little finger to contact the pad of the thumb.
Flexor digiti minimi brevis: Flexes little finger at the metacarpophalangeal joint.

Nerve
Ulnar nerve, C(7), **8**, T**1**.

Basic functional movement
Example: Holding a thread within the fingertips (along with the other fingertips).

Sports that heavily utilise these muscles
Examples: Volleyball. Handball. Rock-face climbing.

Common problems when muscles are chronically tight / shortened
Overabducting (opponens digiti minimi) or overextending (flexor digiti minimi brevis) the little finger resulting from falling into the ulnar side of the hand.

Self stretches

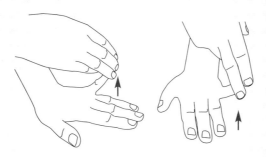

ABDUCTOR POLLICIS BREVIS

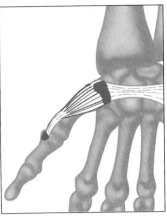

Palmar view, right hand.

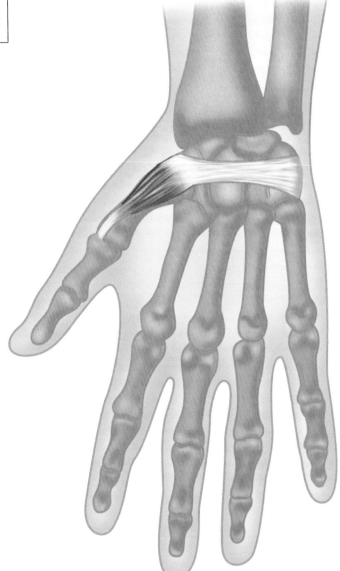

Strengthening exercise

Latin, *abduct*, away from; *pollicis*, of the thumb; *brevis*, short.

This is the most superficial of the muscles of the thenar eminence. The others are flexor pollicis brevis and opponens pollicis.

Origin
Flexor retinaculum. Tubercle of trapezium. Tubercle of scaphoid.

Insertion
Radial side of base of proximal phalanx of thumb.

Action
Abducts thumb and moves it anteriorly (as in typing or playing the piano). Assists in opposition of thumb.

Nerve
Median nerve, (C6, 7, 8, T1).

Basic functional movement
Example: Typing.

Sports that heavily utilise this muscle
Example: Rock-face climbing.

Self stretch

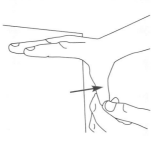

ADDUCTOR POLLICIS

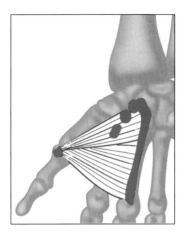

Palmar view, right hand.

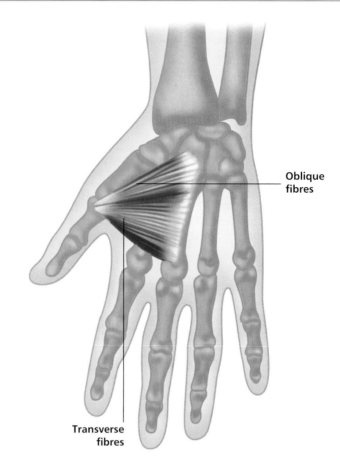

Oblique fibres

Transverse fibres

Latin, *adduct*, toward; *pollicis*, of the thumb.

Origin
Oblique fibres: Anterior surfaces of second and third metacarpals, capitate and trapezoid.
Transverse fibres: Palmar surface of third metacarpal bone.

Insertion
Ulnar (medial) side of base of proximal phalanx of thumb.

Action
Adducts the thumb.

Nerve
Deep ulnar nerve, C8, T1.

Basic functional movement
Example: Gripping a jam jar lid to screw it on.

Sports that heavily utilise this muscle
Example: Rock-face climbing.

Movements or injuries that may damage this muscle
Overabducting the thumb resulting from falling on the hand.

6

Muscles of the Hip and Thigh

GLUTEUS MAXIMUS

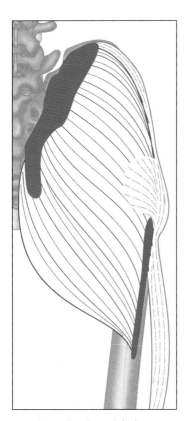

Posterior view, right leg.

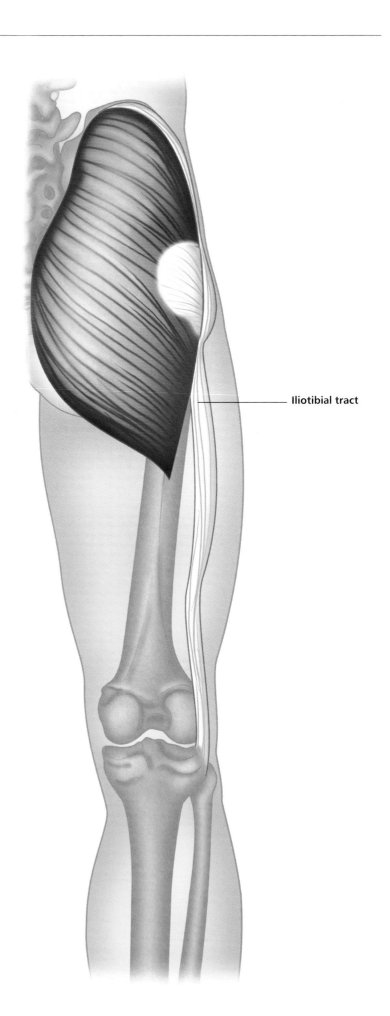

Iliotibial tract

Strengthening exercises

Squats

Seated leg press

Multi-hip machine (cable hip extension / cable kick-back)

Good morning exercise

Greek, *gloutos*, buttocks; *maximus*, biggest.

The gluteus maximus is the most coarsely fibred and heaviest muscle in the body, forming the bulk of the buttock.

Origin
Outer surface of ilium and posterior surface of sacrum and coccyx (over sacroiliac joint).

Insertion
Upper posterior area of femur. Iliotibial tract (long tendon) of fascia lata muscle.

Action
Extends and laterally rotates hip joint (forceful extension as in running or rising from sitting). Extends trunk. Assists in adduction of hip joint.

Nerve
Inferior gluteal nerve, L5, S1, 2.

Basic functional movement
Examples: Walking upstairs. Rising from sitting.

Sports that heavily utilise this muscle
Examples: Running. Surfing. Wind surfing. Jumping. Weightlifting ('clean' phase, i.e. lifting weights up from floor).

Self stretches

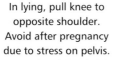

In lying, pull knee to opposite shoulder. Avoid after pregnancy due to stress on pelvis.

TENSOR FASCIAE LATAE

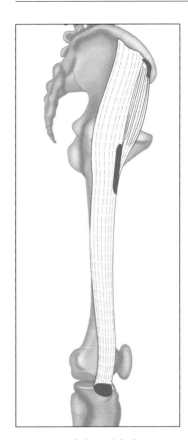

Lateral view, right leg.

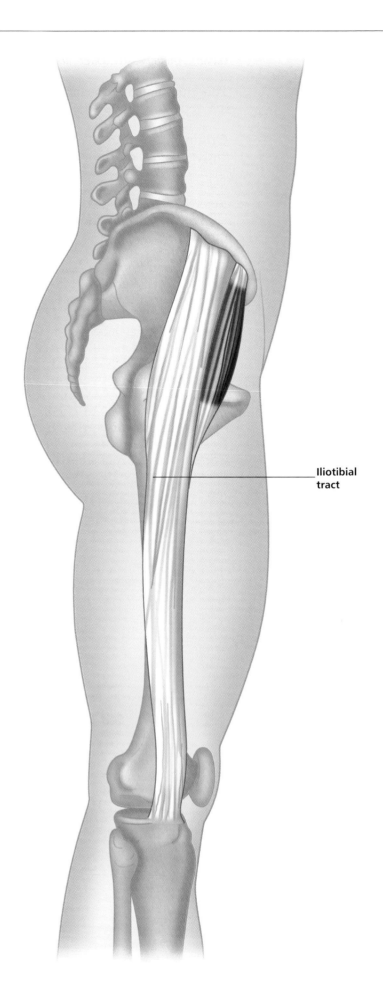

Iliotibial tract

Strengthening exercises

Abductor machine

Multi-hip machine
(cable hip abduction)

Hip abduction

Latin, *tensor*, a stretcher; *fasciae*, band(s); *latae*, broad.

This muscle lies anterior to gluteus maximus, on the lateral side of the hip.

Origin
Outer edge of iliac crest, towards the front.

Insertion
Joins iliotibial tract (long fascia lata tendon) just below the hip, which runs to the upper lateral side of the tibia.

Action
Flexes, abducts and medially rotates the hip joint. Tenses the fascia lata, thus stabilizing the knee.

Nerve
Superior gluteal nerve, L**4**, **5**, S**1**.

Basic functional movement
Example: Walking.

Sports that heavily utilise this muscle
Examples: Horse riding. Hurdling. Water skiing.

Common problems when muscle is chronically tight / shortened
Pelvic imbalances, leading to pain in hips, lower back and lateral area of knees.

Self stretches

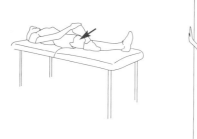

Hand on knee and
pull across body.

Push your hips away
from the wall.

GLUTEUS MEDIUS

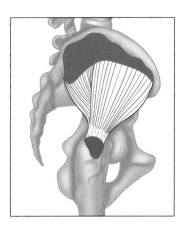

Lateral view, right leg.

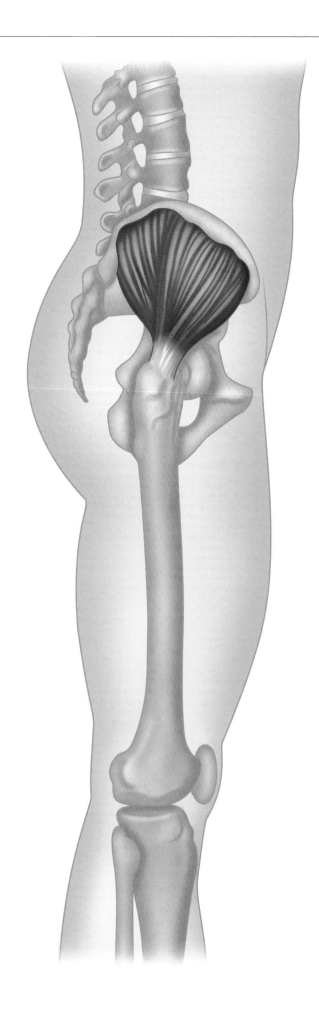

Strengthening exercises

Abductor machine

Multi-hip machine
(cable hip abduction)

Greek, *gloutos*, buttocks; *medius*, middle.

This muscle is mostly deep to and therefore obscured by gluteus maximus, but appears on the surface between gluteus maximus and tensor fasciae latae. During walking, this muscle, with gluteus minimus, prevents the pelvis from dropping towards the non weight-bearing leg.

Origin
Upper outer surface of ilium.

Insertion
Lateral surface of greater trochanter (top) of femur.

Action
Abducts the hip joint. Anterior fibres medially rotate the hip joint. Posterior fibres slightly laterally rotate the hip joint.

Nerve
Superior gluteal nerve, L**4**, **5**, S**1**.

Basic functional movement
Example: Stepping sideways over an object such as a low fence.

Sports that heavily utilise this muscle
Examples: All sports requiring side-stepping, esp. cross-country skiing, ice skating.

Common problems when muscle is chronically tight / shortened
Pelvic imbalances, leading to pain in hips, lower back and knees.

Self stretches

Hand on knee and
pull across body.

Push your hips away
from the wall.

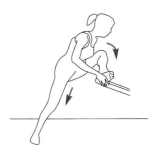

GLUTEUS MINIMUS

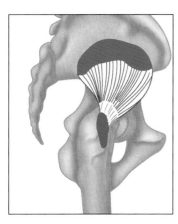

Lateral view, right leg.

Strengthening exercises

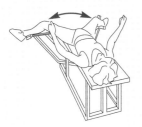

Abductor machine

Multi-hip machine
(cable hip abduction)

Greek, *gloutos*, buttocks; *minimus*, smallest.

This muscle is situated deep to gluteus medius, whose fibres obscure it.

Origin
Middle outer surface of ilium, below origin of gluteus medius.

Insertion
Anterior border of greater trochanter (top) of femur.

Action
Abducts and medially rotates hip joint.

Nerve
Superior gluteal nerve, L**4**, **5**, S**1**.

Basic functional movement
Example: Stepping sideways over an object such as a low fence.

Sports that heavily utilise this muscle
Examples: All sports requiring side-stepping, esp. cross-country skiing, ice skating.

Common problems when muscle is chronically tight / shortened
Pelvic imbalances, leading to pain in hips, lower back and knees.

Self stretches

Hand on knee and
pull across body.

Push your hips away
from the wall.

PIRIFORMIS

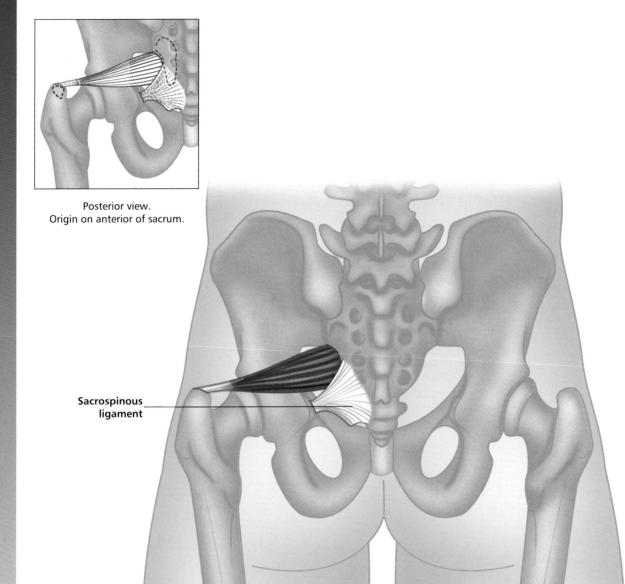

Posterior view.
Origin on anterior of sacrum.

Sacrospinous
ligament

Strengthening exercise

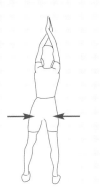

Isometric contraction of the buttocks in standing with legs apart

Latin, *piriform*, pear-shaped; **Greek**, pyramid-shaped.

Piriformis leaves the pelvis by passing through the greater sciatic foramen.

Origin
Internal (front) surface of sacrum.

Insertion
Greater trochanter (top) of femur.

Action
Laterally rotates hip joint. Abducts the thigh when hip is flexed. Helps hold head of femur in its socket.

Nerve
Ventral rami of lumbar nerve, L(5) and sacral nerves, S**1**, **2**.

Basic functional movement
Example: Taking first leg out of car.

Sports that heavily utilise this muscle
Examples: Swimming (breast stroke legs). Soccer.

Common problems when muscle is chronically tight / shortened
Hypertonic muscle may squeeze the sciatic nerve, causing 'piriformis syndrome', i.e. sciatic pain which begins in the buttocks.

Self stretches

Cross right ankle over left knee and bring left knee slowly towards left shoulder, keeping the sacrum in contact with the ground or table. Be careful not to strain your knee joint.

DEEP LATERAL HIP ROTATORS

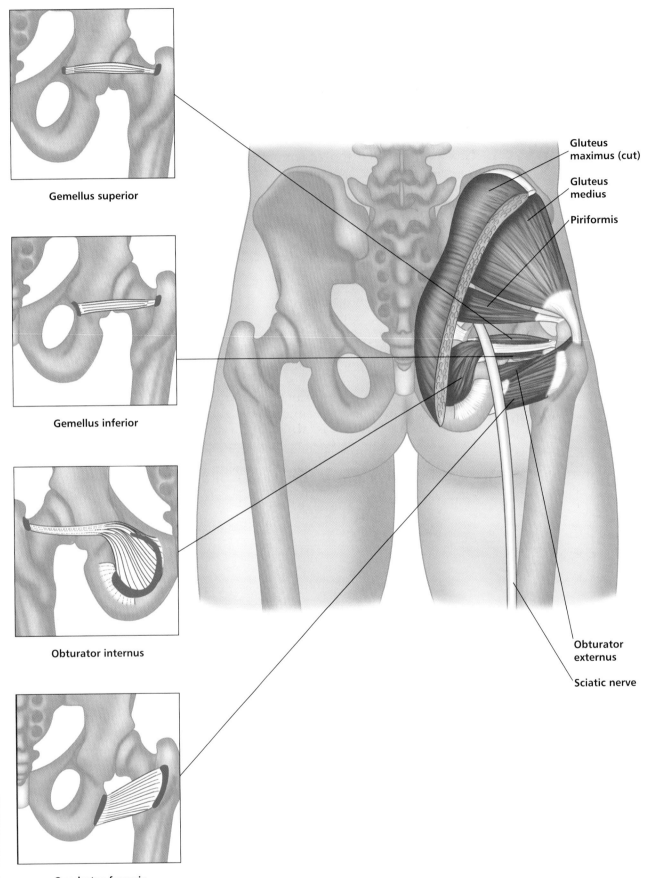

Gemellus superior

Gemellus inferior

Obturator internus

Quadratus femoris

Gluteus maximus (cut)

Gluteus medius

Piriformis

Obturator externus

Sciatic nerve

Includes the obturator internus, the two gemelli, and quadratus femoris. *Gemellus* means "little twin" in **Latin**.

Strengthening exercise

Isometric contraction of the buttocks in standing with legs apart

Origin
Obturator internus: Inner surface of ischium, pubis and ilium.
Gemellus superior: Ischial spine (lower posterior area of pelvis).
Gemellus inferior: Just below origin of gemellus superior.
Quadratus femoris: Lateral edge of ischial tuberosity (sitting bone).

Insertion
Greater trochanter (top) of femur (except quadratus femoris which inserts just behind and below the others).

Action
Laterally rotates hip joint. Helps hold head of femur in its socket (acetabulum).

Nerve
Obturator internus and gemellus superior: Nerve to obturator internus, L**5**, S**1**, **2**.
Gemellus inferior and quadratus femoris: Nerve to quadratus femoris, L**4**, **5**, S**1**, (**2**).

Basic functional movement
Example: Taking first leg out of car.

Sports that heavily utilise these muscles
Examples: Swimming (breast stroke legs). Soccer.

Common problems when muscles are chronically tight / shortened
Person stands with feet turned out.

Self stretch

As for piriformis, although piriformis will receive the most direct stretch from this.

HAMSTRINGS

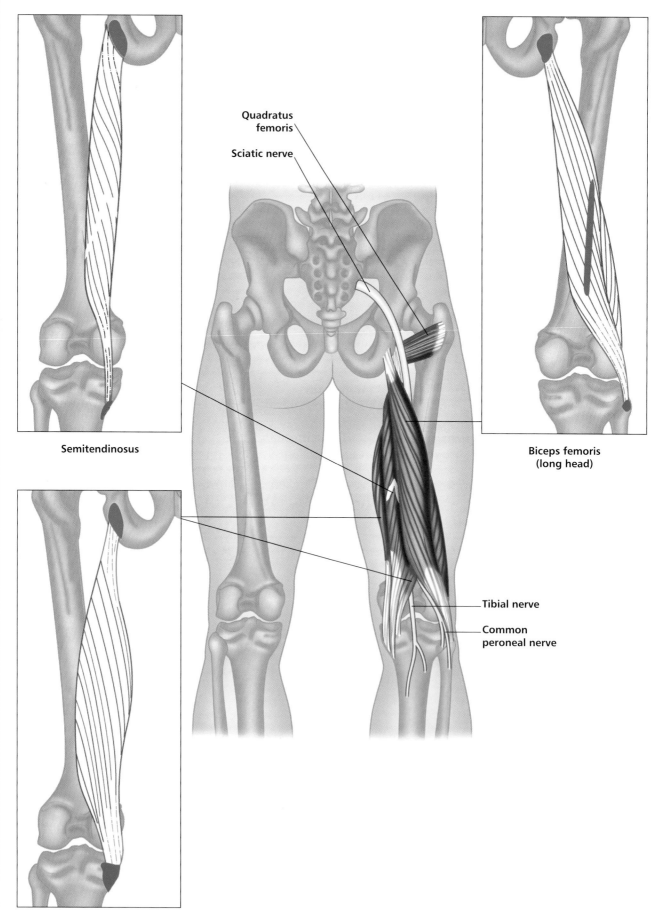

Semitendinosus

Semimembranosus

Quadratus femoris

Sciatic nerve

Biceps femoris (long head)

Tibial nerve

Common peroneal nerve

Strengthening exercises

Leg curl (effects lower portion of hamstrings)

Multi-hip machine (cable hip extension/ kick-back)

Good morning exercise (both effect upper portion of hamstrings)

German, *hamme*, back of leg; **Latin**, *stringere*, draw together.

The hamstrings consist of three muscles. From medial to lateral they are: semimembranosus, semitendinosus and biceps femoris.

Origin
Ischial tuberosity (sitting bone). Biceps femoris also originates from the back of the femur.

Insertion
Semimembranosus: Back of medial condyle of tibia (upper inside part of tibia).
Semitendinosus: Upper medial surface of shaft of tibia.
Biceps femoris: Head (top) of fibula. Lateral condyle of tibia (upper outside part of tibia).

Action
Flex the knee joint. Extend the hip joint.
Semimembranosus and semitendinosus also medially rotate (turn in) the lower leg when knee is flexed. Biceps femoris laterally rotates (turns out) the lower leg when the knee is flexed.

Nerve
Branches of the sciatic nerve, L4, **5**, S1, **2**, 3.

Basic functional movement
During running, the hamstrings slow down the leg at the end of its forward swing and prevent the trunk from flexing at the hip joint.

Sports that heavily utilise these muscles
Examples: Sprinting. Hurdling. Soccer (esp. back kicks). Jumping and weightlifting (upper portion of hamstrings only).

Movements or injuries that may damage these muscles
Sudden lengthening of muscle without sufficient warm-up (e.g. forward kicking, splits).

Common problems when muscles are chronically tight / shortened
Low back pain. Knee pain. Leg length discrepancies. Restriction of stride length in walking or running.

Self stretches

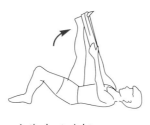

Actively straighten your leg. For tighter hamstrings, hold onto a towel or strap slung over the sole of the foot; or lie in a doorway and fix leg against door jam.

ADDUCTORS

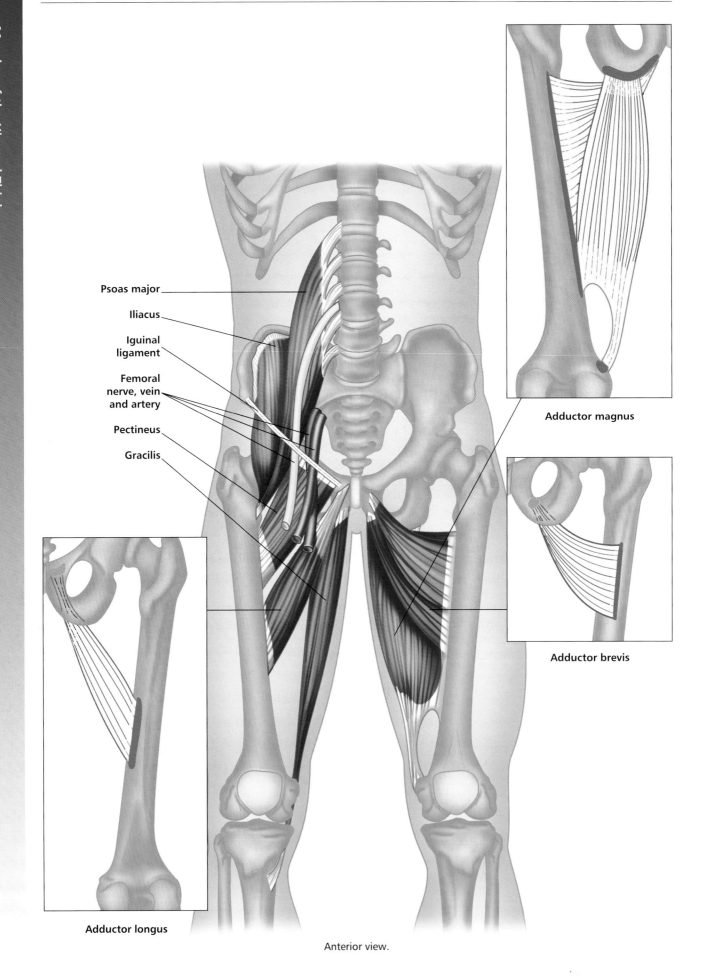

Psoas major

Iliacus

Iguinal ligament

Femoral nerve, vein and artery

Pectineus

Gracilis

Adductor magnus

Adductor brevis

Adductor longus

Anterior view.

Strengthening exercises

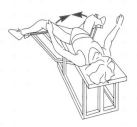

Hip joint adduction machine

Hip adduction

Latin, *adduct*, to bring together.

The adductor magnus is the largest of the adductor muscle group, which also includes adductor brevis and adductor longus. Adductor longus is the most anterior of the three. The lateral border of the upper fibres of adductor longus form the medial border of the **femoral triangle** (sartorius forms the lateral boundary; the inguinal ligament forms the superior boundary).

Origin
Anterior part of pubic bone (ramus). Adductor magnus also takes origin from the ischial tuberosity.

Insertion
Whole length of medial side of femur, from hip to knee.

Action
Adduct and laterally rotate hip joint.
Adductors longus and brevis also flex the extended femur and extend the flexed femur.

Nerve
Magnus: Obturator nerve, L2, **3**, **4**. Sciatic nerve, L**4**, 5, S1.
Brevis: Obturator nerve, (L2–L4).
Longus: Obturator nerve, L**2**, **3**, 4.

Basic functional movement
Example: Bringing second leg in or out of car.

Sports that heavily utilise these muscles
Examples: Horse riding. Judo. Wrestling. Hurdling. Soccer (side passes). Swimming (breast stroke legs). General manoeuvring on court (i.e. crossover steps, side shifting).

Movements or injuries that may damage these muscles
Side splits or high side kicks without sufficient warm-up.

Common problems when muscles are chronically tight / shortened / fatigued
Groin pulls. (The adductors tend to be much tighter in men than in women).

Self stretches

Keep your back straight, with soles of feet together.

GRACILIS

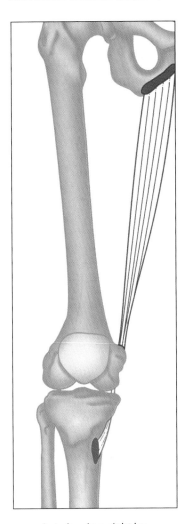

Anterior view, right leg.

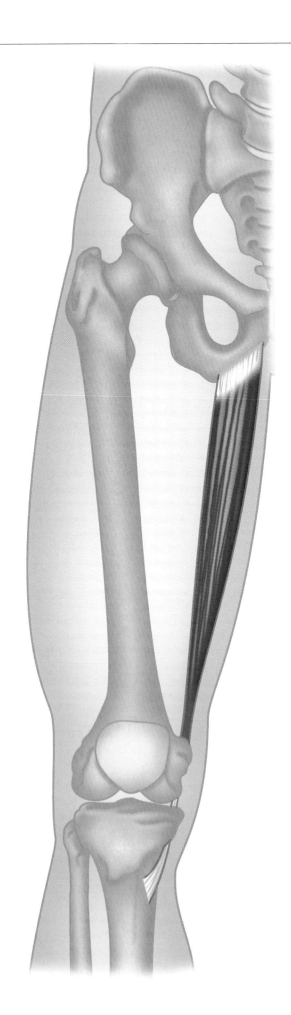

Strengthening exercise

Hip joint
adduction machine

Latin, slender, graceful.

Gracilis descends down the medial side of the thigh in front of semimembranosus.

Origin
Lower margin of pubic bone.

Insertion
Upper part of medial surface of shaft of tibia.

Action
Adducts hip joint. Flexes knee joint. Medially rotates knee joint when flexed.

Nerve
Anterior division of obturator nerve, L**2**, **3**, **4**.

Basic functional movement
Example: Sitting with knees pressed together.

Sports that heavily utilise this muscle
Examples: Horse riding. Hurdling. Soccer.

Movements or injuries that may damage this muscle
Side splits or high side kicks without sufficient warm-up.

Common problems when muscle is chronically tight / shortened / fatigued
Groin pulls. (The adductors tend to be much tighter in men than in women).

Self stretches

Keep your back
straight, with soles
of feet together.

PECTINEUS

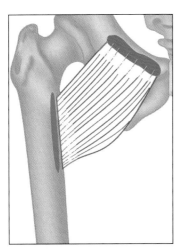

Anterior view, right leg.

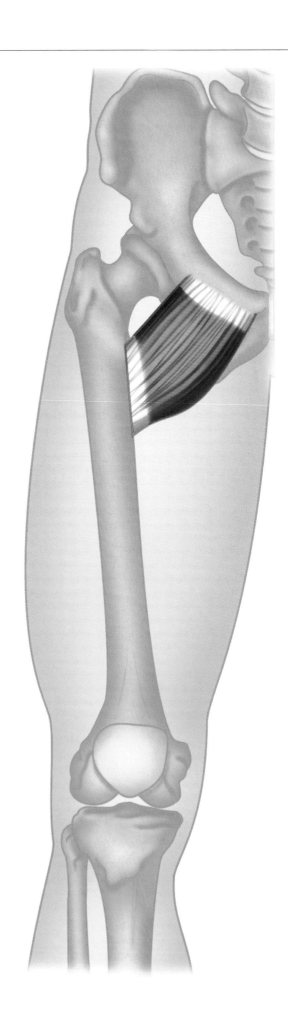

Strengthening exercises

Hip joint adduction machine

Multi-hip machine (cable hip flexion)

Hanging leg raise

Latin, *pecten*, comb; *pectenate*, shaped like a comb.

Pectineus is sandwiched between the psoas major and adductor longus.

Origin
Upper anterior (superior ramus) area of pubic bone.

Insertion
Upper medial shaft of femur.

Action
Adducts the hip joint. Flexes the hip joint.

Nerve
Femoral nerve, L**2**, **3**, 4. Occasionally receives an additional branch from the obturator nerve, L3.

Basic functional movement
Example: Walking along a straight line.

Sports that heavily utilise this muscle
Examples: Horse riding. Rugby. Sprinting (maximizes stride length). Kicking sports (e.g. soccer, to maximise kicking force).

Movements or injuries that may damage this muscle
Side splits or high side kicks without sufficient warm-up.

Common problems when muscle is chronically tight / shortened / fatigued
Groin pulls. (The adductors tend to be much tighter in men than in women).

Self stretches

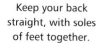

Keep your back straight, with soles of feet together.

SARTORIUS

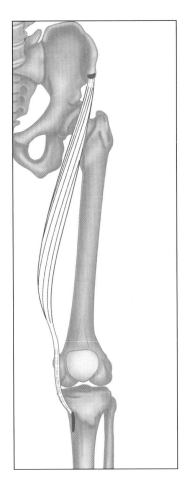

Anterior view.

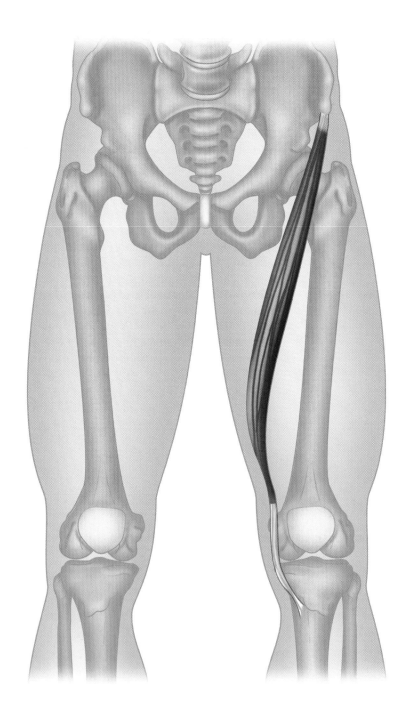

Strengthening exercise

Multi-hip machine
(cable hip abduction)

Sartorius is the most superficial muscle of the anterior thigh. The medial border of the upper third of this muscle forms the lateral boundary of the **femoral triangle** (adductor longus forms the medial boundary; the inguinal ligament forms the superior boundary). *Sartorius* is so named because it means "tailor" in **Latin**; and its action is to put the lower limbs in the cross-legged seated position of the tailor.

Origin
Anterior superior iliac spine (i.e. the most anterior point of the ilium).

Insertion
Upper part of medial surface of tibia.

Action
Flexes hip joint (helping to bring leg forward in walking or running). Laterally rotates and abducts the hip joint. Flexes knee joint. Assists in medial rotation of the tibia on the femur after flexion. These actions may be summarized by saying that it places the heel on the knee of the opposite limb.

Nerve
Two branches from the femoral nerve, L**2**, **3**, (**4**).

Basic functional movement
Example: Sitting cross-legged.

Sports that heavily utilise this muscle
Examples: Ballet. Skating. Soccer.

Movements or injuries that may damage this muscle
Being over ambitious with yoga exercises in cross-legged or lotus position (although the knee is likely to be damaged first).

Common problems when muscle is chronically tight / shortened
Pain or damage to inside of the knee.

Self stretches

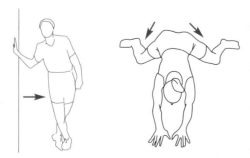

Push your hips away
from the wall.
Only a slight stretch.

QUADRICEPS

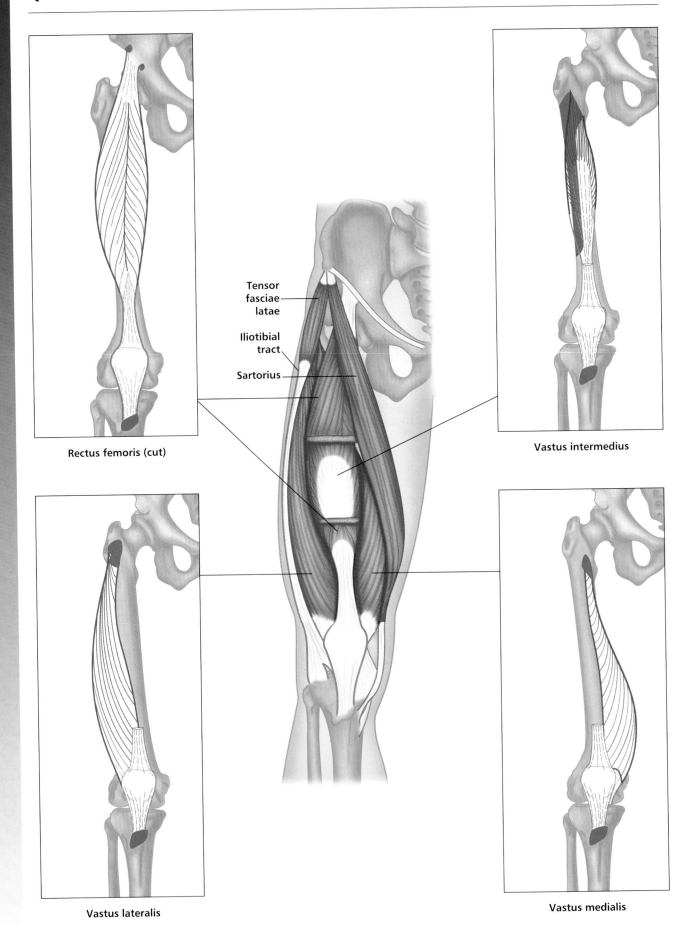

Rectus femoris (cut)

Tensor
fasciae
latae

Iliotibial
tract

Sartorius

Vastus intermedius

Vastus lateralis

Vastus medialis

Strengthening exercises

Quads-knee extension

Leg press

Squats

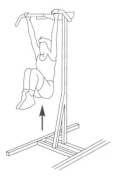

For rectus femoris only:
Hanging leg raise

For rectus femoris only:
Multi-hip machine
(cable hip flexion)

Latin, four-headed; **Greek**, four-footed.

The four quadriceps muscles are: rectus femoris, vastus lateralis, vastus medialis, and vastus intermedius. They all cross the knee joint, but the rectus femoris is the only one with two heads of origin and that also crosses the hip joint. The quadriceps straighten the knee when rising from sitting, during walking and climbing. The vasti muscles as a group pay out to control the movement of sitting down.

Origin
Rectus femoris: Front part of ilium (anterior inferior iliac spine). Area above hip socket.
Vastus group: Upper half of shaft of femur.

Insertion
Patella, then via patellar ligament into the upper anterior part of the tibia (tibial tuberosity).

Action
Vasti: Extends the knee joint.
Rectus femoris: Extends the knee joint and flexes the hip joint (particularly in combination, as in kicking a ball).

Nerve
Femoral nerve, L**2**, **3**, **4**.

Basic functional movement
Example: Walking up stairs. Cycling.

Sports that heavily utilise these muscles
Examples: Fell running (push off phase and knee stability when running). Skiing. All jump events. Kicking sports (soccer, karate, etc.). Weight lifting.

Common problems when muscles are chronically tight / shortened
Low back pain. Knee pain, knee instability; esp. if tight and weak.

Self stretches

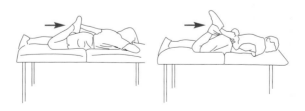

Use opposite hand to hold your ankle,
or wrap a towel around
your leg and use both hands.

7

Muscles of the Leg and Foot

TIBIALIS ANTERIOR

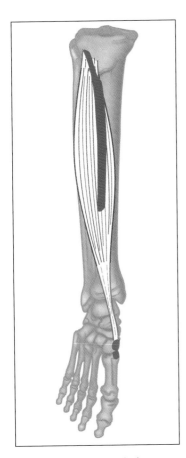

Anterior view, right leg.

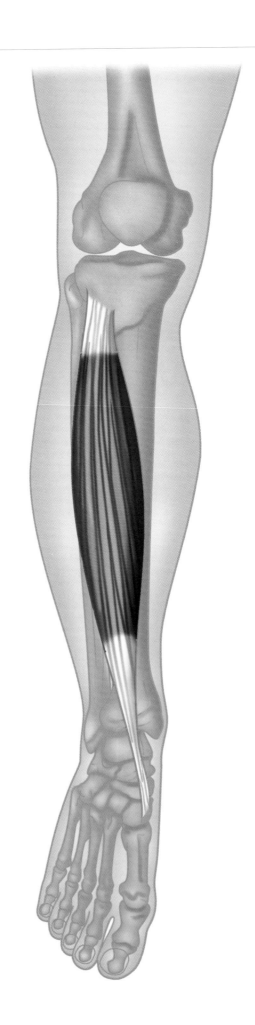

Strengthening exercises

Strengthening exercises

Toe raise

Quads knee extension

Latin, *tibia*, pipe or flute, shinbone; *anterior*, before.

Origin
Upper half of lateral and anterior surface of tibia (including lateral condyle of tibia).

Insertion
Medial edge of front of foot (medial cuneiform bone and base of first metatarsal).

Action
Dorsiflexes (lifts up) the foot. Inverts the foot.

Nerve
Deep peroneal nerve, L4, **5**, S1.

Basic functional movement
Example: Walking and running (helps prevent the foot from slapping onto the ground after the heel strikes. Lifts the foot clear of the ground as the leg swings forward).

Sports that heavily utilise this muscle
Examples: Hill walking. Mountaineering. Running. Breast stroke swimming. Cycling (the pedal up phase).

Movements or injuries that may damage this muscle
Excessive jumping onto hard surfaces.

Self stretches

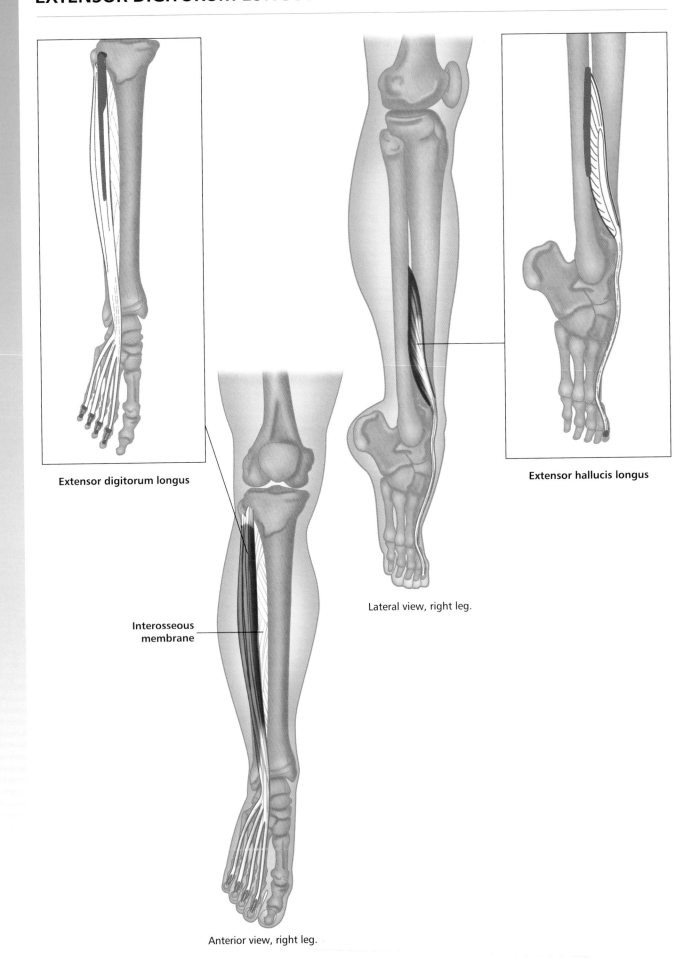

Extensor digitorum longus

Extensor hallucis longus

Interosseous membrane

Lateral view, right leg.

Anterior view, right leg.

Strengthening exercise

Toe raise

Latin, *extensor*, to extend; *digit*, toe; *hallux*, great toe; *longus*, long.

Like the corresponding tendons in the hand, the extensor digitorum longus forms extensor hoods on the dorsum of the proximal phalanges of the foot. These hoods are joined by the tendons of the lumbricales and extensor digitorum brevis, but not by the interossei. The extensor hallucis longus lies between and deep to tibialis anterior and extensor digitorum longus.

Origin
Anterior surface of fibula and interosseous membrane (fibrous tissue between the tibia and fibula). Extensor digitorum longus also arises from the lateral condyle (upper outer part) of tibia.

Insertion
Extensor digitorum longus: Phalanges of lateral four lateral toes.
Extensor hallucis longus: Distal phalanx of great toe.

Action
Extensor digitorum longus: Extends toes. Dorsiflexes ankle joint and everts foot.
Extensor hallucis longus: Extends great toe. Dorsiflexes ankle joint and inverts foot.

Nerve
Fibular (peroneal) nerve, L4, **5**, S1.

Basic functional movement
Walking up the stairs (ensuring the toes clear the steps).

Sports that heavily utilise these muscles
Examples: Hill walking. Mountaineering. Breast stroke swimming. Cycling (the pedal up phase).

Movements or injuries that may damage these muscles
Tendon easily bruised by compression (e.g. if toe is stepped on).

Self stretches

FIBULARIS (PERONEUS) LONGUS AND BREVIS

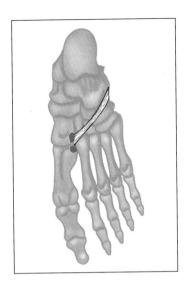

Plantar view, right leg.

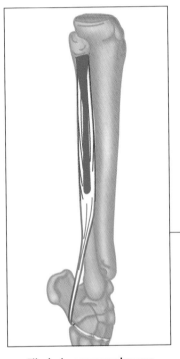

Fibularis peroneus longus

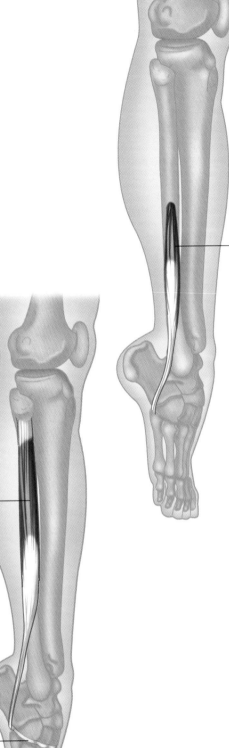

Fibularis (peroneus) longus tendon (seen through tarsal bones)

Lateral view, right leg.

Fibularis peroneus brevis

Strengthening exercises

Calf raise
(standing heel raise)

Calf raise
(seated heel raise)

Latin, *fibula*, pin / buckle; *longus*, long; *brevis*, short.

The course of the tendon of insertion of fibularis longus helps maintain the transverse and lateral longitudinal arches of the foot. A slip of muscle from fibularis brevis often joins the long extensor tendon of the little toe, whereupon it is known as *peroneus digiti minimi*. Fibularis tertius is a partially separated lower lateral part of extensor digitorum longus.

Origin
Longus: Upper two-thirds of lateral surface of fibula.
Brevis: Lower two-thirds of lateral surface of fibula.

Insertion
Longus: Base of first metatarsal.
Brevis: Base of fifth metatarsal.

Action
Everts the foot. Assists plantar flexion of ankle joint (i.e. points the foot).

Nerve
Superficial fibular (peroneal) nerve, L4, **5**, S**1**.

Basic functional movement
Example: Walking on uneven surfaces.

Sports that heavily utilise these muscles
Examples: Running. Soccer. Jumping.

Movements or injuries that may damage these muscles
Forced inversion of the ankle (i.e. overstretching the lateral aspect of the ankle) may create chronic problems with ankle joint stability.

Self stretches

Stretch increases
dorsiflexion rather
than inversion.

GASTROCNEMIUS

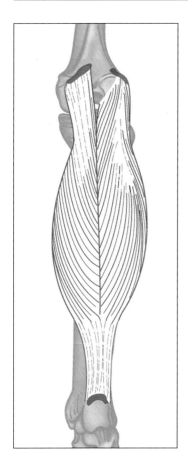

Posterior view, right leg.

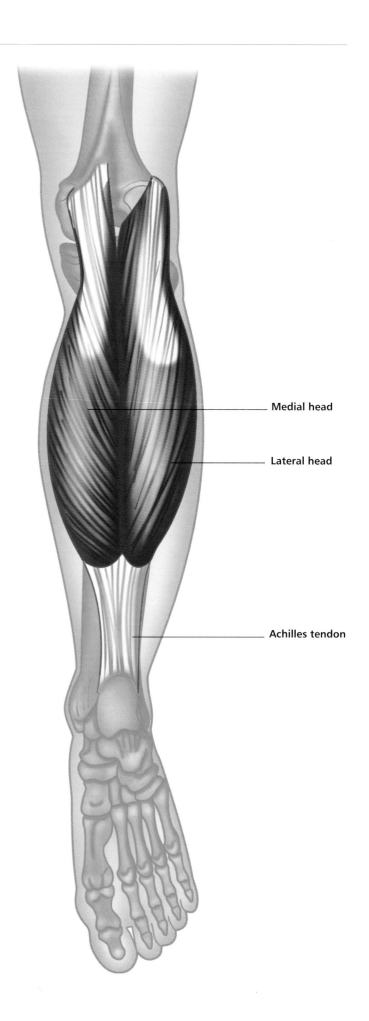

Medial head

Lateral head

Achilles tendon

Strengthening exercises

Calf raise
(standing heel raise)

Leg curl

Greek, *gaster*, stomach; *kneme*, leg.

Gastrocnemius is part of the composite muscle known as **triceps surae**, which forms the prominent contour of the calf. The triceps surae comprises: gastrocnemius, soleus and plantaris. The **popliteal fossa** at the back of the knee is formed inferiorly by the bellies of gastrocnemius and plantaris, laterally by the tendon of biceps femoris, and medially by the tendons of semimembranosus and semitendinosus.

Origin
Medial head: Lower posterior surface of femur above medial condyle.
Lateral head: Lateral condyle and lower posterior surface of femur.

Insertion
Posterior surface of calcaneus (heel bone) via the calcaneal tendon (Achilles tendon); which is a fusion of the tendons of gastrocnemius and soleus.

Action
Plantar flexes (points) foot at ankle joint. Assists in flexion of knee joint. It is a main propelling force in walking and running.

Nerve
Tibial nerve, S**1**, **2**.

Basic functional movement
Standing on 'tip-toes'.

Sports that heavily utilise this muscle
Examples: Most sports requiring running and jumping, esp. sprinting, high jump, long jump, volleyball, basketball. Ballet. Push off in the swim start. Trampoline.

Movements or injuries that may damage this muscle
Explosive jumping, or landing badly when jumping down, may rupture the tendocalcaneous (Achilles tendon) at its junction with the muscle belly.

Common problems when muscle is chronically tight / shortened
Constant wearing of high-heeled shoes tends to cause this muscle to shorten, which can effect postural integrity.

Self stretches

Stretch increases
dorsiflexion rather
than inversion.

SOLEUS

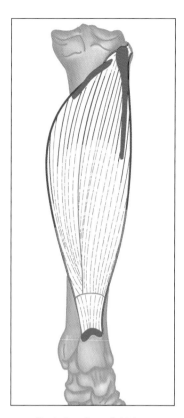

Posterior view, right leg.

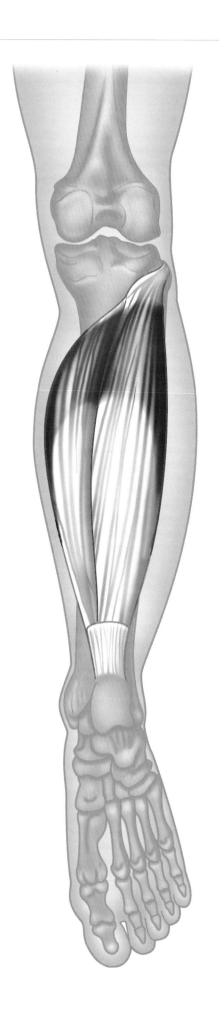

Strengthening exercises

Calf raise
(standing heel raise)

Calf raise
(seated heel raise)

Latin, sole-shaped (fish).

Part of the **triceps surae**. The calcaneal tendon of the soleus and gastrocnemius is the thickest and strongest tendon in the body.

Origin
Upper posterior surfaces of tibia and fibula.

Insertion
With gastrocnemius via calcaneal tendon into posterior surface of calcaneus (heel bone).

Action
Plantar flexes ankle joint. The soleus is frequently in contraction during standing to prevent the body falling forwards at the ankle joint; i.e. to offset the line of pull through the body's centre of gravity. Thus, it helps to maintain the upright posture.

Nerve
Tibial nerve, L5, S**1**, **2**.

Basic functional movement
Standing on 'tip-toes'.

Sports that heavily utilise this muscle
Examples: Most sports requiring running and jumping, esp. sprinting, high jump, long jump, volleyball, basketball. Ballet. Push off in the swim start. Trampoline.

Movements or injuries that may damage this muscle
Explosive jumping, or landing badly when jumping down, may rupture the tendocalcaneous (Achilles tendon) at its junction with the muscle belly.

Common problems when muscle is chronically tight / shortened
Tight and painful calves or tendocalcaneous (which is usually more a problem of soleus than gastrocnemius). Constant wearing of high-heeled shoes tends to cause this muscle to shorten, which can affect postural integrity.

Self stretches

Stretch increases
dorsiflexion rather
than inversion.

POPLITEUS

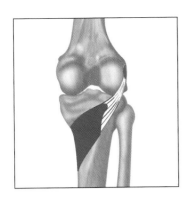

Posterior view, right leg.

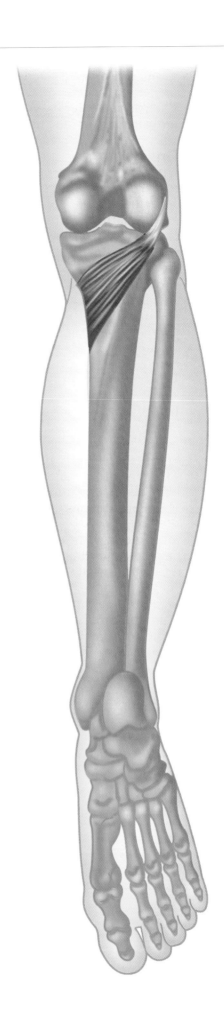

Strengthening exercise

Leg curl

Latin, *poples*, ham.

The tendon from the origin of popliteus lies inside the capsule of the knee joint.

Origin
Lateral surface of lateral condyle of femur. Oblique popliteal ligament of knee joint.

Insertion
Upper part of posterior surface of tibia, superior to soleal line.

Action
Laterally rotates femur on tibia when foot is fixed on the ground. Medially rotates tibia on femur when the leg is non-weight bearing. Assists flexion of knee joint, (popliteus 'unlocks' the extended knee joint to initiate flexion of the leg). Helps reinforce posterior ligaments of knee joint.

Nerve
Tibial nerve, L4, **5**, S1.

Basic functional movement
Example: Walking.

Sports that heavily utilise this muscle
All activities involving running and walking.

Movements or injuries that may damage this muscle
High kicks without sufficient warm-up.

Common problems when muscle is chronically tight / shortened
Inability to fully extend knee joint, possibly resulting in knee pain or injury.

TIBIALIS POSTERIOR

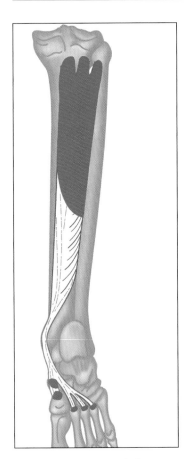

Posterior view, right leg.

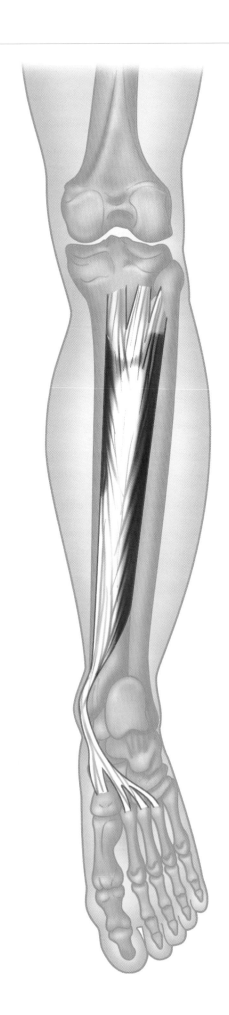

Strengthening exercises

Calf raise
(standing heel raise)

Calf raise
(seated heel raise)

Latin, *tibia*, pipe / flute, shinbone; *posterior*, behind.

Tibialis posterior is the deepest muscle on the back of the leg. It helps maintain the arches of the foot.

Origin
Posterior surface tibia and fibula, and most of the interosseous membrane.

Insertion
Tarsal bones (navicular, cuneiforms, cuboid, sustentaculum tali of calcaneus), and the second, third and fourth metatarsals.

Action
Inverts the foot. Assists in plantar flexion of the ankle joint.

Nerve
Tibial nerve, L(4), **5**, S**1**.

Basic functional movement
Standing on 'tip-toes'. Pushing down car pedals.

Sports that heavily utilise this muscle
Examples: Sprinting. Long jump. Triple jump.

Movements or injuries that may damage this muscle
Poor alignment of the lower limb, especially walking or standing with feet turned out, will cause collapse of the medial longitudinal arch of the foot.

Self stretches

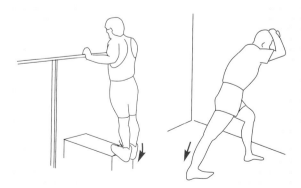

FLEXOR DIGITORUM LONGUS

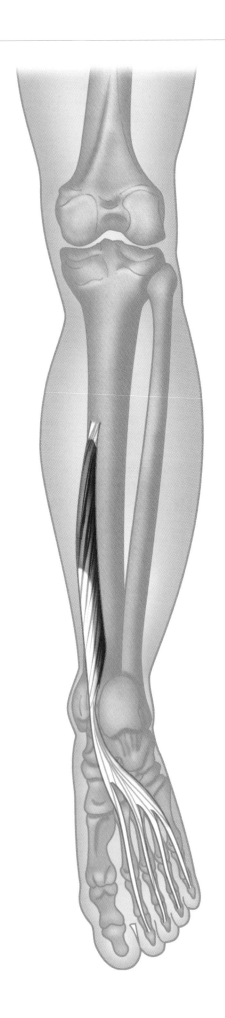

Posterior view, right leg.

Strengthening exercises

Calf raise
(standing heel raise)

Calf raise
(seated heel raise)

Latin, *flex,* to bend; *digit,* toe; *longus,* long.

The insertion of the tendons of this muscle into the lateral four toes parallels the insertion of flexor digitorum profundus in the hand.

Origin
Medial part of posterior surface of tibia.

Insertion
Distal phalanges of second through fifth toes.

Action
Flexes all the joints of the lateral four toes (enabling the foot to firmly grip the ground when walking). Helps to plantar flex the ankle joint and invert the foot.

Nerve
Tibial nerve, L5, S1, (2).

Basic functional movement
Walking (esp. bare foot on uneven ground). Standing on tip-toes.

Sports that heavily utilise this muscle
Examples: Ballet. Gymnastics (beam work). Karate (side kick).

Common problems when muscle is chronically tight / shortened
Hammer toe deformity of lateral four toes.

Self stretches

Curl your toes under to extend the toe joints.

Keep your toes extended.

FLEXOR HALLUCIS LONGUS

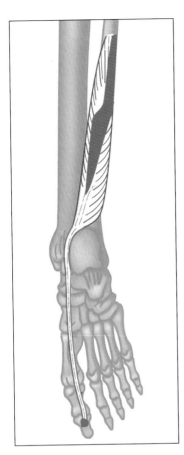

Posterior view, right leg.

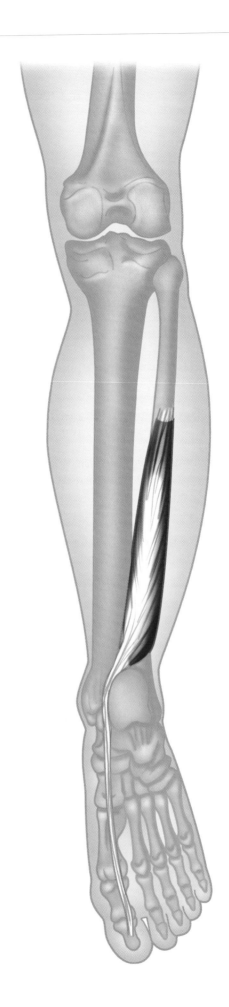

Strengthening exercises

Calf raise
(standing heel raise)

Calf raise
(seated heel raise)

Latin, *flex*, to bend; *hallux*, great toe; *longus*, long.

This muscle helps maintain the medial longitudinal arch of the foot.

Origin
Lower two-thirds of posterior surface of fibula. Interosseous membrane.

Insertion
Distal phalanx of great toe.

Action
Flexes the great toe. Helps to plantar flex and invert the foot. Helps stabilize the inside of the ankle.

Nerve
Tibial nerve, L5, S1, **2**.

Basic functional movement
Pushing off the surface in walking (esp. bare foot on uneven ground). Standing on 'tip-toes'.

Sports that heavily utilise this muscle
Examples: Running. Hill walking. Ballet. Gymnastics.

Common problems when muscle is chronically tight / shortened
Hammer toe deformity of great toe.

Self stretches

Curl your toes under to
extend the toe joints.

Keep your toes extended,
especially the big toe.

LUMBRICALES

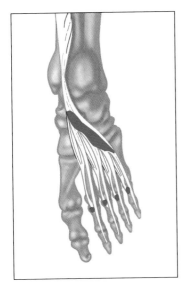

Plantar view, right foot.

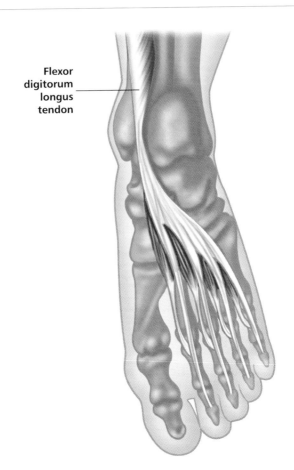

Flexor digitorum longus tendon

Latin, earthworm.

Origin
Tendons of flexor digitorum longus.

Insertion
Medial side of base of proximal phalanges of second through to fifth toes and corresponding extensor expansion.

Action
Flex the metatarsophalangeal joints and extend the interphalangeal joints of the lateral four toes.

Nerve
Lateral three lumbricales: Lateral plantar nerve, L(4), (5), S1, **2**.
First lumbricalis: Medial plantar nerve, L4, **5**, S1.

Basic functional movement
Example: Gathering up material under the foot using the toes only.

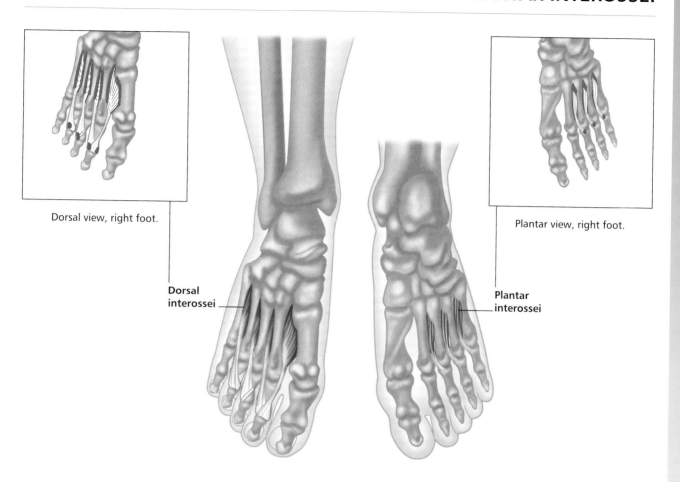

Dorsal view, right foot.

Plantar view, right foot.

Dorsal interossei

Plantar interossei

Latin, *dorsum*, back; *plantar*, sole of the foot; *interosseus*, between bones.

Similar to the hand, the dorsal interossei are larger than the plantar interossei.

Origin
Dorsal interossei: Adjacent sides of metatarsal bones.
Plantar interossei: Bases and medial sides of third, fourth and fifth metatarsals.

Insertion
Dorsal interossei: Bases of proximal phalanges:
First: Medial side of proximal phalanx of second toe.
Second to fourth: Lateral sides of proximal phalanges of second to fourth toes.
Plantar interossei: Medial sides of bases of proximal phalanges of same toes.

Action
Dorsal interossei: Abduct (spread) toes. Flex metatarsophalangeal joints.
Plantar interossei: Adduct (close together) toes. Flex metatarsophalangeal joints.

Nerve
Lateral plantar nerve, S**1**, **2**.

Basic functional movement
Example: Facilitates walking.

Sport that heavily utilises these muscles
Running, especially with bare feet.

Resources

Alter, M. J.: 1998. *Sport Stretch: 311 Stretches for 41 Sports*. Human Kinetics, Champaign.

Biel, A.: 2001. *Trail Guide to the Body, 2nd edition*. Books of Discovery, Boulder.

Clemente, C. M. (editor): 1985. *Gray's Anatomy of the Human Body, 30th edition*. Lea & Febiger, Philadelphia.

Cook, B. B., and Stewart, G. W.: 1996. *Strength Basics*. Human Kinetics, Champaign.

DeJong, R. N.: 1967. *The Neurological Examination, 2nd & 3rd editions*. Harper & Row, New York.

Foerster, O., and Bumke, O.: 1936. *Handbuch der Neurologie (vol. V)*. Publisher unknown, Breslau.

Haymaker, W., and Woodhall, B.: 1953. *Peripheral Nerve Injuries, 2nd edition*. W. B. Saunders Co., Philadelphia.

Jarmey, C.: 2004. *The Atlas of Musculo-skeletal Anatomy*. Lotus Publishing/North Atlantic Books, Chichester/Berkeley.

Kendall, F. P., and McCreary, E. K.: 1983. *Muscles, Testing & Function, 3rd edition*. Williams & Wilkins, Baltimore.

Lycholat, T.: 1990. *Complete Book of Stretching*. Crowood Press, Marlborough.

McAtee, R. E., and Charland, C.: 1999. *Facilitated Stretching, 2nd edition*. Human Kinetics, Champaign.

Norris, C. M.: 1997. *Abdominal Training*. A & C Black, London.

Norris, C. M.: 1999. *The Complete Guide to Stretching*. A & C Black, London.

Norris, C. M.: 1993. *Weight Training Principles & Practice*. A & C Black, London.

Romanes, G. J. (editor): 1972. *Cunningham's Textbook of Anatomy, 11th edition*. Oxford University Press, London.

Schade, J. P.: 1966. *The Peripheral Nervous System*. Elsevier, New York.

Spalteholz, W.: (date unknown). *Hand Atlas of Human Anatomy (vols. II & III, 6th edition)*. J. B. Lippincott, London.

Tortora, G.: 1989. *Principles of Human Anatomy, 5th edition*. Harper & Row, New York.

Yessis, M.: 1992. *Kinesiology of Exercise*. Masters Press, Lincolnwood.

General Index

Index of Muscles